# Dilation of Time and Space

*An Examination of the True Nature of Spacetime*

by

## Miroslav Halza

# Contents

# Preface

Over a century ago, Albert Einstein—as a philosopher and theoretical physicist—studied the relationship of physical observations to speed. His logical analyses of certain deviations he observed led him to create his Special Theory of Relativity, where he presented these deviations as "time dilation." Basically, he said, time slows for an object as its speed increases. This is one of the reasons why the term "relativity" comes into play. Time varies relative to speed; it's not a perfect one-to-one relationship. Or at least, it isn't according to our identity.

But then, we human beings are imperfect. Our main defect, when it comes to understanding the universe, is our delayed response to the natural world due to the processing of incoming information about natural motion. Geometrical and mathematical analysis of experimental observations should be compared to physical laws; when philosophy is added, the physical theory is not always applicable to real physics.

Einsteinian geometry and mathematics have resulted in a number of new physical theories. For instance, Einsteinian physics suggest that when clocks act differently under certain physical conditions, time itself changes under those conditions. One example is gravitational time dilation, when time slows due to increased gravity. Before Einstein, physicists would have considered that the changed time dimension was caused by a new physical environment (i.e., another operating frame) for the clock. Another example is that the old physics suggests that the lifetime of a particle in one frame of reference may not be the same as it is in another frame. But modern physics says that the lifetime *must* be the same, independently of the changed physical conditions. Therefore, time itself changes character, rather than the particle; hence, time dilation.

By my way of thinking, Einstein's philosophy describes how human beings perceive nature and nothing more. In reality, the physical world is independent of human activities and perception. That is why physics must be stripped of human additions contradictory to physical laws, and applied *only* to data acquired by experimental physics. One example of a human addition is the idea that a particle in another reference frame that does not maintain the laws of mass and energy (and thus is not from our world) determines the physical properties (i.e. mass) of particles in our material world.

One branch of physics, String Theory, postulates that the smallest possible unit of physical matter is a one-dimensional string. String Theory can be used to explain gravitation and time deviations without resorting to ideas like time or space dilation or contraction. Although many theorists believe they can use String Theory to understand all the forces and particles in nature, it's a very poor String Theory that does. They use only the mathematical formulae for *some* string properties, and therefore their String Theory cannot actually be the ultimate physical theory of fundamental particles and forces. Genuine string theorists, in their physical approach to strings, must work with **all the physical variants of strings**.

The string theory of music is better described physically, and better understood, than the existing String Theory of physics. In this report, I hope to improve this state by exploring how String Theory applies to both temporal and spatial distortion—apparent dilation and contraction. This is founded on a basic understanding of elementary strings; the shapes of strings propagating in space; the effects that free strings cause in nature; how strings are bound into subatomic particles; the shapes of strings that exist on the surfaces of subatomic particles; and which forces produce collisions of strings having momentum, and so on. In this treatise, the explanations will be brief, since a detailed

discussion of the basics was published in my previous book, *Absolute String Theory*.

## Elementary Strings

Small units are the building blocks of larger structures. At the most basic level, everything in our world consists of sets of chemical compounds built from atoms. Atoms are themselves comprised of nuclei, composed of protons and neutrons, orbited by electrons. Nature consists not just of matter, but also of energy and the forces arising from the interactions between objects. The universe registers these interactions between distant objects, and therefore there must also exist particles that create the forces across these distances. Clearly, there are particles *smaller* than the subatomic particles that comprise atoms. According to String Theory, the smallest particles take the forms of elementary strings.

## The Definition of a String

The smallest object possible is a point object. Strings consist of connected points, forming a line. The points of this line are not fixed in relationship to each other; therefore, variations in length and in shapes exist among these elementary strings. Their flexibility can cause a string to be curved or straighten, prolonged or shortened. Thus, a string is a segment of the smallest possible elastic (live) curve. Here, the term "live" means that the string bears the smallest amount of energy possible in nature— the basic unit or quantum of energy. Its "life" lies in its ability to move: to vibrate back and forth as a macroscopic spring does, to rotate upon itself (spinning), or to move its loose ends. This flexibility allows for the existence of strings in many dynamic geometrical forms.

## Strings Propagating through Space

Vibration is an intrinsic property of freely existing strings in nature. Since physics allows two kinds of vibrations relative to space, two kinds of vibrating strings must exist. Vibration must be considered specially for moving strings, because they propagate very quickly and therefore do not have enough time to adjust themselves to the needs of their surroundings, which is why they maintain their inner motion.

Moving strings propagating at the speed of light vibrate either longitudinally (forward and backward) and transversely (side to side). **Transversely vibrating strings are photons.** We know photons very well, since the existence of all life is based on the action of photons. Among other things, photons create light, making the statement "we are beings of light" literally true.

All other strings propagating at the speed of light must vibrate longitudinally, making them gravitons (see the details of my theoretical argument in *Absolute String Theory*). We don't registering gravitons directly, but we can detect their effects, suggesting that these statements about them should be true:

1. Since nature allows two kinds of vibrations and one of them is well known, then strings vibrating in the other direction—longitudinally—must propagate through space the same way photons do, and at the same speed.

2. Although we are beings of light, the natural world exists thanks to the force that binds matter together—gravity. The gravitational force is so dominant in the universe that many scientists deny the existence of any other large-scale universal forces (they think gravitation is the all-mighty force of the universe left after the Big Bang). Physics teaches that a force is just an effect of interacting bodies; to teach otherwise is

to teach pseudoscience or mysticism. Hence, gravitation *must* be the result of interacting strings that propagate through space, though not as transverse waves. Since gravitation affects bodies located quite far apart in the universe, these strings must propagate at the speed of light. If gravitation exists, then **longitudinally vibrating strings propagating at the speed of light** must exist. They do, and are called gravitons.

## Strings in Matter

Bound strings form matter. This has been confirmed by experimental observation of strings (photons) emitted from subatomic particles of matter. Therefore, the elemental unit of subatomic particles is the string. Some researchers believe that quarks are the elemental particles of matter, but they fail to realize that quarks also emit strings directly or as products of their decay. The conclusion? That strings must be the smallest elemental particles in ordinary matter. Strings existing inside subatomic particles mostly have the same form, and give matter its basic physical property: mass.

Since I have explained mass in detail in my previous publications, for now I'll just comment that mass is the physical property of matter arising from the dynamics of strings inside matter. "Dynamic" means that an object moves; and therefore, there must be room for movement inside particles of matter. Since a string has the intrinsic property of movement (its life), this is enough to prove that there is plenty of free space inside subatomic particles to allow strings to move. If a baryon (a proton or neutron) has a great deal of free space in its volume, then it should be possible to squeeze baryonic matter into a smaller volume given enough pressure.

Extreme pressure would be able to squeeze baryonic matter into a very small volume indeed; and theoretically, it can. We have nothing that can compress matter this way, since it would be impossible to do so using equipment built from ordinary matter, i.e., matter that is based on baryonic matter. However, the extreme pressure present when an especially large star dies triggers the further gravitational collapse of matter. That gravitational pressure removes any free space in atoms, creating a very dense form of neutron matter, or *neutronium*, that collapses into what astrophysicists call a neutron star. But matter can be compressed even farther, so much that the density at the hearts of the largest dying stars produces black holes.

Yet we live in an industrial age, in which technicians have been able to construct a machine on Earth where the extreme pressure needed for such a collapse can also be reached, for a very short time, during particle collisions. The Large Hadron Collider at CERN speeds up baryons almost to the speed of light, and then collides them head on. The LHC reached such an extreme colliding pressure on July 3, 2012 that observers detected a piece of matter with a density many times higher than that of normal baryons. They reported that this particle lived for $1.56 \times 10^{-22}$ seconds.

The lifetime of this particle was equal to the amount of time it took light to propagate the distance of $4.68 \times 10^{-14}$ m. Hence, the collision ran at or near the speed of light, since it lasted just as long in light meters as the diameter of a nucleus (a cluster of baryons). This type of particle can exist in our world only under extreme conditions, under extreme pressure, and only for as long as that collision time. When the collision ended, this particle decayed back into ordinary matter (fermions and other particles) and strings (mostly photons). The conclusion here is that the latest experiments at LHC in

CERN created a piece of matter proving the existence of **substantial free space in the volume of subatomic particles.**

The scientific implication is that the elemental particles comprising subatomic particles are similar in their volumes to the molecules of a gas. The smallest particle known to nature is the string, and therefore, strings should move within the volumes of subatomic particles. By general assumption, their movement is random, like the Brownian movement of gases. Therefore, I conclude: **the mass of an elementary subatomic particle arises from the dynamics of the strings comprising its volume**. The LHC's latest experiments have proved this theory of mass.

**Gravitons Interacting with Matter**

Gravitons are traveling strings vibrating longitudinally, propagating through space at the speed of light. But strings residing in matter (i.e., in subatomic particles) are moving as well, so we need to resolve which kind of vibration prevails among the strings comprising matter. If gravitons interact with matter, then they must be the same kind of strings; and so the same kind of vibration should occur in both. Therefore, I conclude that the strings filling the volume of subatomic particle also vibrate longitudinally.

If two strings are vibrating in the same direction and they meet, let's say at the same velocity, they should interfere physically, as springs do, and pair up.

Vibrating strings relocating in subatomic particles should collide randomly with neighboring strings (per Brownian movement), and therefore do not have a uniform and one-directional motion. Their movement should be like that of an object connected to a spring, or like an object suspended from a pivot, swinging freely as a pendulum. In either case, the string accelerates as it approaches equilibrium (the midpoint position of its movement), then slows down until it stops at the point of maximal displacement, and then starts to accelerate back toward equilibrium, slowing again until it reaches the opposite point of maximal displacement. This means that there must be a higher speed, at least at the equilibrium point, than the speed of light, which we observe in strings propagating through space (Newton's Third Law).

Gravitation starts to work when a graviton merges in a "sticky collision" with a string of a particle of matter, and then both accelerate to the equilibrium point. After passing the equilibrium point, the pair of strings slows until their speed is equal to the graviton's initial speed (the speed of light). As the speed of the couplet tries to fall below the normal speed of light, the graviton is forced to leave the string of matter (in order to maintain its identity). The graviton continues on until it eventually (and temporarily) merges with another string of matter. When that string of matter loses the graviton, the string continues on to the point of maximal displacement, then turns back—and so on.

From the above description of gravitons merging with strings of matter, it's clear that strings of matter give gravitons "rides" during their motion. The rides have these physical consequences:

1. The gravitons are displaced through space farther and faster than they would be without the interactions, and therefore, gravitons accelerate as they propagate through matter. **Gravitons propagate faster through matter than they do through a vacuum**—which means they move through matter faster than the speed of light. Thus, the index of refraction for gravitons propagating through matter is always less than 1,

contrary to the index of refraction for photons (light). The index of refraction = the velocity of strings in a vacuum/the velocity of strings propagating through a medium.

2. If gravitons displace farther when propagating though matter, then they gain speed in the direction of propagation. According to the physical law of action and reaction (Newton's Third Law of Motion), a string giving a graviton a ride then has to recoil; resulting in a speed less than its former speed without the ride. The graviton captures some of the linear momentum of the string of matter during the sticky collision and subsequent ride. Since the strings that the gravitons propagate through reside in subatomic particles, the subatomic particles are pushed in the direction of the graviton emitter as the strings recoil. Thus, **gravitons push matter back in the direction from which they came** (per Newton's Law of Universal Gravitation).

3. When the strings of matter slow down because the resulting speed of the sticky collision, this is the result of the sum of both linear momenta from the graviton and the string of matter. If the host string slows during the transfer, then it needs more time to reach maximal displacement in order to eliminate this lost momentum. This also means that when an atomic clock "ticks" in relation to the motion of its subatomic particles, gravitation must also affect it (a topic we will discuss later in more detail). Therefore, we should not assume that gravity has slowed time. We do not need new physics to explain this, since physics already explains the slower "ticking" in a high-gravity field. **Gravitational time dilation does not exist in string theory physics.**

**Entropy at the Spontaneous Birth of Gravitons**

Strings in matter move, and therefore have dynamic properties. The dynamic properties of objects are not quantitatively stable forever, but change in directions necessary to ease internal stress (per the Second Law of Thermodynamics). Since every string in a subatomic particles moves, it has kinetic energy. The maximal kinetic energy of the string occurs at its highest speed—at the equilibrium point of its range of movement. At the point of maximal displacement, when the string stops, its kinetic energy is zero. Its kinetic energy changes into potential energy; and thus the maximal potential energy is at the maximal displacement point.

The constant change of potential energy into kinetic energy and back results in stress within the subatomic particle. The fact that the strings are locked within the subatomic particles creates natural stress. However, a subatomic particle can lose some of this internal energy (stress) when a string escapes from it.

A freed string should travel at the speed of light, and thus carry away its kinetic energy. The string acquired that kinetic energy somewhere between its equilibrium point and its maximal displacement point, when it had a speed equal to the speed of light. However, the string also had some potential energy as well as kinetic energy at that point. When a string escapes from a particle, this potential energy is lost to a decrease in order in the universe.

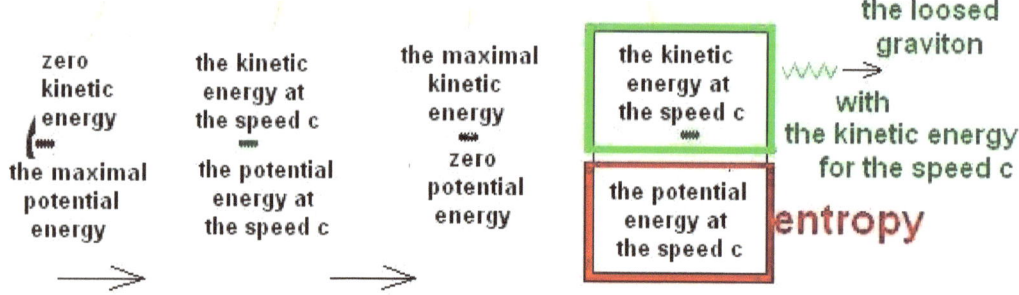

This lost energy is the smallest quantum of entropy in the universe. It is lost to nature; though it's not destroyed, it goes nowhere, so it can never be reused. This means that our material world was created or evolved with added energy in the past, but now benefits from the creator's or evolution's energy as applied to the previously discussed strings in their restricted volume. In a sense, then, our material world "burns" this energy to sustain its life and functioning. This lost energy is the engine of all processes in the universe. Spontaneous universal processes (stellar creation, galaxy formation, the creation of chemical elements in stars and stellar explosions, the production of energy in stars, and so on) all exploit this creative potential. In science, this entropy creates spontaneous action in nature.

The spontaneous birth of gravitons from ordinary matter must arise due to the Second Law of Thermodynamics applying to matter having mass. Consequently, spontaneous action in the macroworld (our normal-sized world) runs in the direction where gravitons are emitted more often (water always flows downhill, and a small object moves in the direction of the massive object).

Every object made of ordinary matter emits gravitons. Their rate of emission is proportional to the stress inherent in matter, and thus is proportional to the mass of the body. Newton's Law of Universal Gravitation reflects this, because it states that a gravitational force between two bodies is directly proportional to their masses. The force generated by gravitons is an attractive force toward another object. Since an object emits gravitons into space in all directions, their density falls by the square of the distance from the object, so the gravitational force is disproportional to the square of the distance. **This is a physical explanation of Newton's Law of Universal Gravitation, according to String Theory.**

Therefore, Einstein's conjecture that gravitation is due to the bending of space due to the presence of matter is not borne out in modern String Physics.

## Strings for Electricity

Strings inside a particle of matter should vibrate longitudinally (back and forth), like springs. If we were to see them on the surfaces of subatomic particles, some would resemble needles, giving the particles a hedgehog-like appearance. The surface strings have one end rooted inside the particle, while the other is free to stretch outside the particle.

A string with one end free might wave around, due to its longitudinal motion. If the string is also rotating around its length, then the waving end creates a funnel shape and, thus, a vortex: ⊳━━━ . As occurs in macroworld whirlpools, the conical mouth of such a **funnel** might drag in other things. So, funnel strings have the ability to attract stem strings, and stem strings have the ability to be so attracted.

We should expect that both shapes of strings will exist on the surfaces of most particles, so both forces (pushing into and pulling in) are present. Thus, individual strings will have either a positive or a negative charge, based on their shapes. Nonetheless, a particle as a whole may have a surface with an overall neutral charge when all positive and negative charges are summed. That particle is a neutron. Now, remember that although the neutron's *overall* charge is neutral, its individual strings have different charges, which stresses the surface. Each charged particle wants to be paired with another of the opposite charge. This stress causes instability, which is why free neutrons decay spontaneously, with a half-life of about ten minutes. Otherwise, when a neutron touches another neutron, the charged longing between some surface strings is satisfied when they form pairs, causing the neutrons to bond. This generates the strong nuclear force between the neutrons, and therefore, neutrons in nuclei are stable. Their half-life is extended indefinitely.

Free neutrons decay into positively and negatively charged particles. The largest of these should be the one with the greatest "hedgehog" appearance; that is, the one that has stems extending beyond the surface. This is the positively charged proton. Each of the surface stems bears a tiny positive charge.

Conversely, particles with a majority of open vortices on the surface have a negative charge. Each little vortex has its own tiny negative charge. When summed with the charges of all the other strings, this results in a negatively charged particle—an electron.

When surface strings move actively enough on the neutron, they may break free and be emitted. As some of these strings start to emerge, they pull loose other strings fixed to them (mass's strings inside the neutron), and thus a new particle comes into existence that has a net negative surface charge—i.e., an electron. Since this process of decay occurs without the influence of collisions with other particles, this new product may not travel away from the baryon, which now has a positive charge and is thus a proton. The decay products may instead become bound to form an atom. In an atom, the force of the linear momentum of the electron is choked off by the electrical force, so the electron acquires rotational momentum and orbits the proton indefinitely. This is the elemental ordinary matter.

Note that the neutron precedes the atom and its other constituents. Hence, at the beginning of time, the universe must have gone through a stage when only neutrons existed.

**Strings for Magnetism**

Strings without fixed ends may also exist among the myriads of strings in a subatomic particle. The ends of such a string may attach to each other, forming a circle that can't spin normally, so it rotates like a wheel instead. These "wheel strings" have a physical property whereby one attracts another rotating in the opposite direction. If two wheels having the same axis of rotation rotate in the same direction, they repel each other. This is the origin of magnetism.

As mentioned in the previous section, when an electron forms from a neutron, it disturbs or breaks a whole string system in the baryon, also casting out strings giving mass to the electron. New single-string wheels are also emitted. When such a string leaves the neutron, it may pair up with another that circles in the opposite direction, while that wheel takes along another, and so on. The result can be a much larger particle consisting of conjoined, alternately rotating wheel strings—a magnetic particle.

This particle lacks an electrical charge, because it possesses no electrical strings. Therefore, it can depart the decaying particle without any electric restrictions. Similarly, because gravity affects only vibrating strings, it does not affect these wheel strings. Therefore, these new particles can propagate at the speed of light without gravitational restrictions. However, if this particle also pulls out some mass's strings (vibrating longitudinally), then it has a mass and can be affected by gravitational fields. Science calls these particles neutrinos.

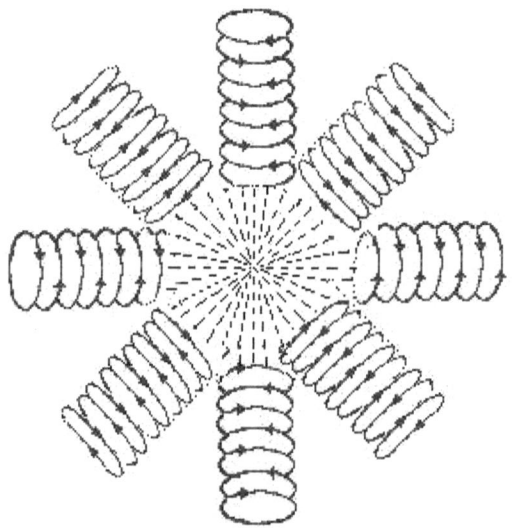

Standard neutron decay produces a particle with a majority of positively charged stemmed strings on its surface, a **proton**, as well as a particle with a majority of negatively charged funnel-shaped strings on its surface, an **electron**. Together, they may form a hydrogen atom, which is the most common type in the universe (not surprising if the universe began as a burst of neutrons). Also produced are wheel-shaped strings, **neutrinos**, that may or may not have a very tiny mass. These three particles are all stable elementary particles spontaneously created by nature.

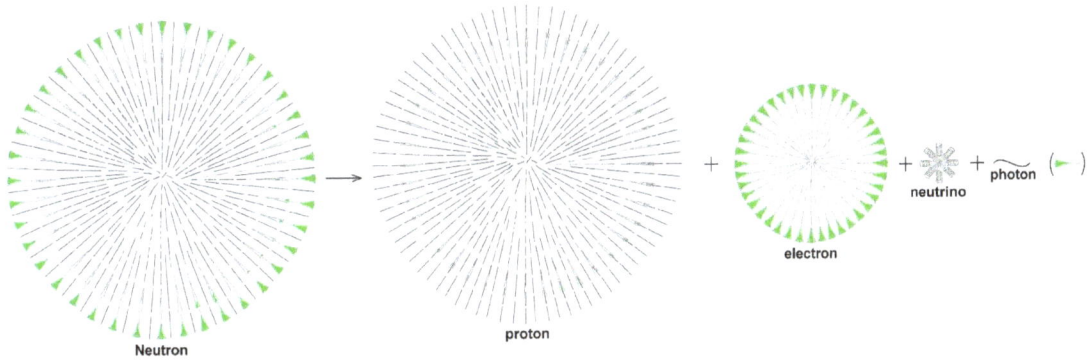

## The Nuclear Forces

The strong nuclear force acts only inside the nuclei of atoms; hence its name. It's this force that holds protons and neutrons together. It arises between particles when their strong electric charges come into contact and "short circuit"—that is, when funnel-shaped surface strings of one particle hold onto the stems of other particles. The strong nuclear force works only over very, very short distances. Consider the two neutrons and two protons bond into the nucleus of an atom of helium:

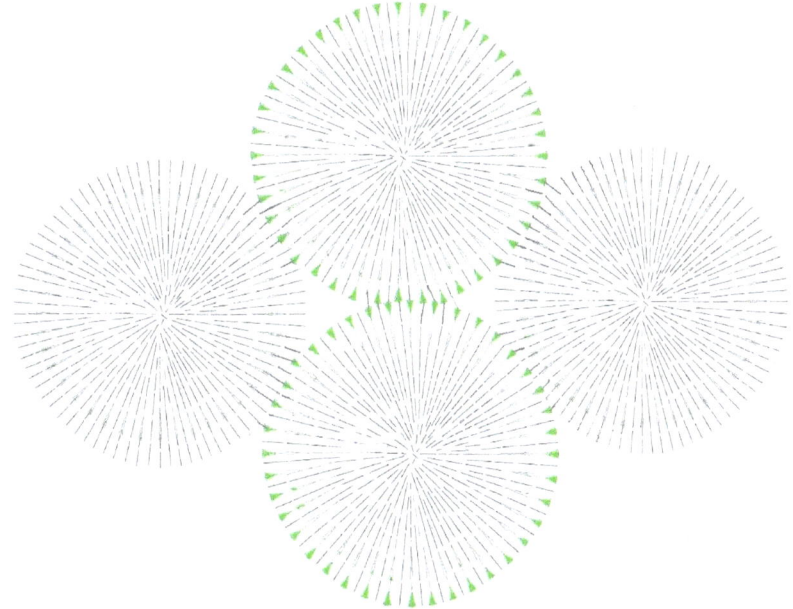

Since protons and neutrons are spherical, it's possible to multiply the "plugging joints" of nuclear particles into many varieties. The forces required to hold a nucleus together may vary. In some cases, a large cluster of baryons, consisting of many poorly arranged neutrons and protons, can solve this stress by decomposing into smaller clusters—that is, into new nuclei. Science explains this as a result of the **weak nuclear force**.

## Strings for Electric and Magnetic Field Lines

Although some strings propagate at the speed of light—the standard speed for string movement—logically, some are not propagating through space at all. These strings lack momentum, and therefore don't carry force to other objects. But they're still "live;" each possesses a quantum of energy. This allows them to adjust themselves to other such strings very quickly by changing their shapes. These strings have no fixed ends, and can form shapes without any restrictions. Thus, they can become the building blocks for connecting channels between electric charges and magnetic poles. Rather than having just one funnel shaped end, they can have two.

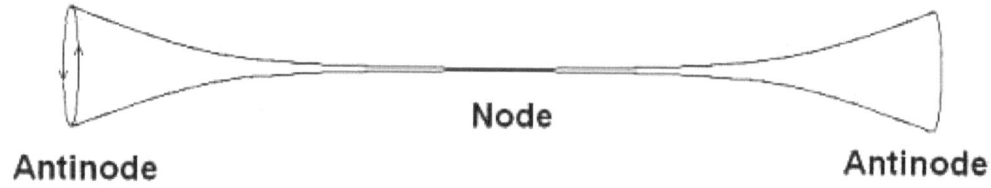

**Node**

**Antinode**          **Antinode**

These strings can shape themselves to fit their circumstances. For instance, when they approach a string with a negative electrical charge, they can adjust themselves so that one end is a positive stem to plug into the approaching vortex, and the other becomes a negatively charged vortex itself. Other neighboring non-propagating strings then adjust to the previous string, so that a stem is pointed to the vortex of that string, and a new vortex is created on opposite end of the second string, and so on—thus forming a channel that seeks a positively charged source. This forms an electric field. The charge is satisfied when it finds a positively charged object, or finds another end of a chain that has built up from a firm positive charge. In this way, an electrical field line is formed and completed between two charges.

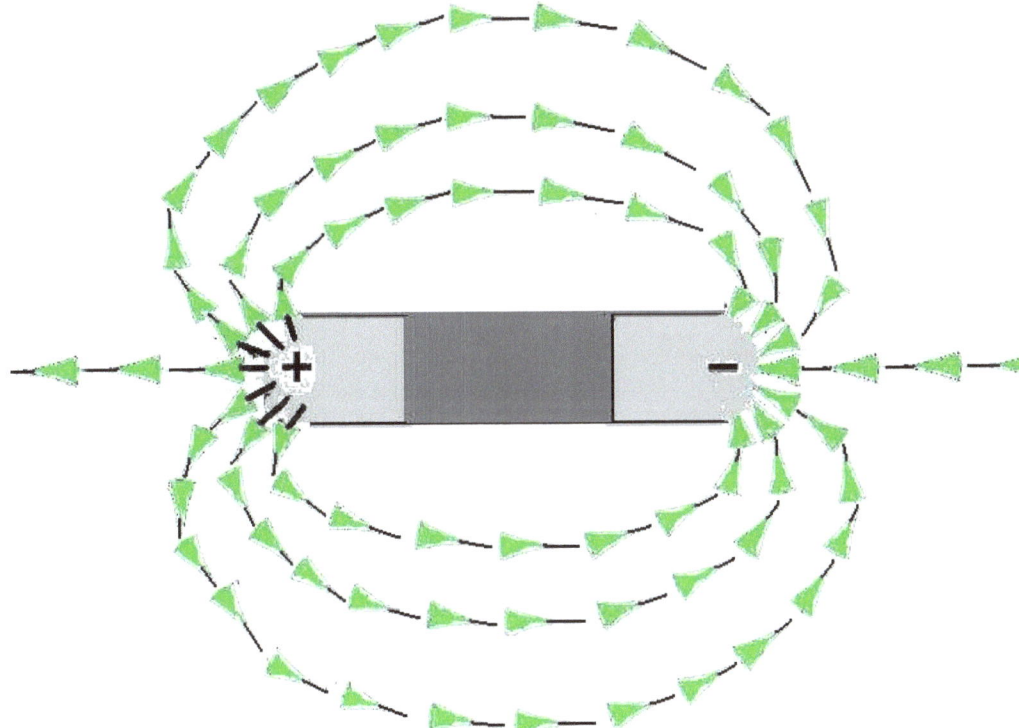

If the charges on objects are not fixed, then they can move from one to the other thanks to this line of force. There is an attractive force between electric charges, since our strings seek to join one another. Certainly, as the stems more closely approach the vortexes, there arise stronger effects on the stems; and therefore, the force between two charges increases as the distance between them decreases.

In fact, more lines can be created between two charges, mostly depending on the numbers of charged strings on the surface of the electrical contacts. A more highly charged electric contact allows free strings to create more lines, and therefore the force between electric charges depends also on their electrical potentials (Coulomb's law).

Sometimes both electric charges can run in the same direction. In this case, incomplete chains approach vortex-to-vortex or stem to stem. This sets up a repulsive electric force between the same charges (positive to positive or negative to negative).

The same holds true at magnetic poles. Free strings there adjust themselves to rotate differently from neighboring strings. In this way, they create magnetic field lines. Where are magnetic field lines there is a magnetic field. Then electric field lines create an electric field. Thus, strings without momenta (without **c**) carry local electric and magnetic forces among electric and magnetic charges.

**Electromagnetism in Photons**

Force is an effect between interacting objects. Since the smallest subatomic objects are strings, the basic interactions that generate forces in nature arise from the interactions of these strings with matter. Since there are just two ways for strings to act on distant objects, we usually define just two fundamental forces between objects outside the atomic nucleus. We have two carriers of forces, therefore: two bosons, one each for gravity (the graviton, which possesses longitudinal vibration) and electromagnetism (the photon, with its transverse vibration). The question is, why do photons apparently carry

two forces (electricity and magnetism) in one quantum of energy?

Photons usually come into existence when they escape from the surfaces of

electrons. The electron loses a surface string, which propagates along a straight line.

However, this string is not a straightened abscissa, since funnel-shaped strings cover the

surface of the electron, due to the strings having one end rooted inside the electron and

the second free. Therefore, it has a node at the fixed end and an antinode (the funnel

shape) in the free end. When such a string pulls loose as a photon, this string has both

ends free, and therefore the node displaces toward the middle of the string. Such a string

propagates through space as a photon, which is constantly spinning along its long axis.

Therefore, the trajectory of photons is not a one-dimensional line, but more of a three-

dimensional spiral. This explains how a photon can have characteristics of both a particle

and a wave.

As the photon propagates through space, it can be modeled as uniform circular

motion combined with perpendicular uniform velocity, as pictured.

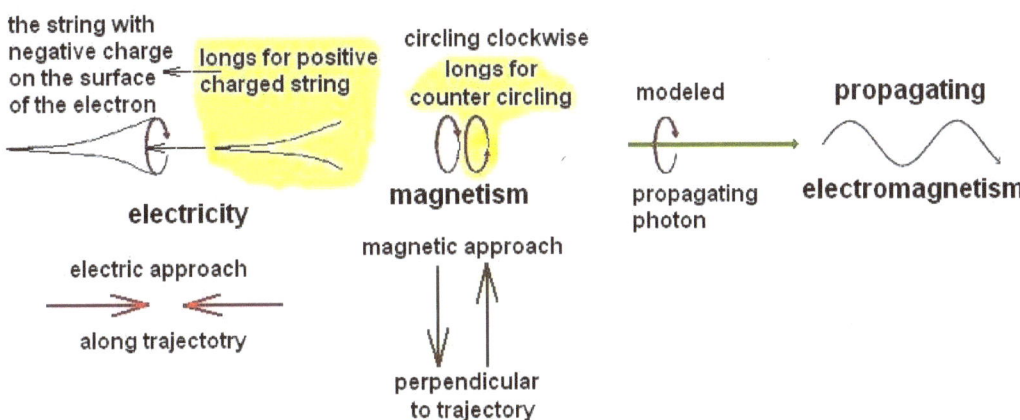

Earlier, I described magnetism as a force arising from the rotation of a wheel

string around its imaginary axis; the magnetism is an effect of that rotation. You can see that the modeled photon is circling, and therefore the photon also has the properties of rotation. Hence the magnetism in "electromagnetism."

A force can either pull or push, so a force has magnitude and direction. To distinguish how many forces are present, we must also look at the direction of the forces. If there's one source and one direction, then there's one force. However, even if there's only one source but two directions, then there are two forces. Now, let's look at the physical substance of the photon again. The negative electrical charge of the photon vortex seeks a positive charge **in the direction of propagation.** The magnetic charge of the photon (the modeled circular motion) seeks another "wheel" rotating in the opposite axis of rotation, **perpendicular to the electrical force.** Therefore, we see two natural forces embodied in one boson.

When a photon is absorbed by an electron, the electron gets one more string on its surface. Since the electron bears mostly funnel-shaped strings, our photon becomes a funnel-shaped string and adds its electric potential to the electron. The magnetic potential is harder to predict, since the magnetic potential of the string is caused by the rotation of the wheel shape. To do that, our string should be disconnected from the electron to form a wheel.

The magnetic force is also present in electrons; and there it is not due to the physical effect of the component strings, but due to the physical effect of the entire particle. The electron itself rotates around its own axis, even as it revolves around the nucleus. This spinning gives rise to the magnetism of the electron. An addition to this, spin might be observed as increased magnetism.

When an object moves, it possesses a speed. If the object has a speed, then it has momentum. An object propagating along a straight line has linear momentum. Since the string is the elemental object of the universe, then a string like our photon, propagating at the speed of light, has linear momentum. That means that when the photon collides with another object in an elastic collision, the conservation of momentum determines the effects of this collision. We've already seen what happens when gravitons collide with other strings of subatomic particles and transfer the gravitational force.

Strings are invisible to us; whereas we can see matter at a macroscopic scale. But why do we see matter at all? Because our eyes register the photons as incoming light. If an object is in the way of light, we see this object, because light from the object is not coming into our eye, or at least that does is less intense. Now, photons mostly stop when they hit non-transparent objects, due to interference with the electrons orbiting atomic nuclei in the objects. The electrons absorb the photons in a "sticky collision" that transfers momentum to the object. In accordance with Newton's Third Law, this generates a force. When electrons absorb photons, they have absorbed their momentum. Yet the electron orbiting the nucleus has both angular orbital momentum and angular spinning momentum.

Now, when the electron is struck by the photon (and does not produce the photoelectric effect, in which the electron is eject from the atom), then the linear momentum of the photon is split partly into angular momenta and partly into a linear momentum of the whole electron, in a way that moves the whole atom and moves the electron into a higher orbit. Then the absorbed photon increases the orbital and spin momenta of the electron. Hence, the magnetic effect of the electron (the spinning) would increase, due to the linear momentum of the photon, rather than the electromagnetism of the photon.

This gives the atom more linear momentum, which we may feel as added heat. If the matter consists only of the simplest atoms (i.e., hydrogen), then it should be no surprise that a cloud of hydrogen can be pushed away by an object emitting photons. Indeed, this is a common occurrence in deep space. The acting force causes acceleration according to Newton's Second Law. If the collision force caused by photons prevails over the gravitational force of a photonic emitter, then even a very low rate of acceleration lasting for long periods can give the colliding particles a significant speed.

Theoretical physicists (if they still want to be called physicists), must account for *all* of the effects caused by photons—the photon's electric effect, its magnetic effect, and its linear momentum effect. With the resulting collision force, adding the linear momentum to atoms must always push them in direction toward which the photons propagate, away from the photonic emitter. Therefore, theoretic physicists shouldn't make the claim that repulsion of objects is an unknown physical effect of dark energy, since natural physics already describes the repulsion of objects in the universe very well.

# TIME

## The Life of Strings in Time

Energy is a necessary property of all living things. It gives them life. Life is related to light, so in a sense, we are all beings of light. Light is made up of energy quanta called photons, and photons carry information. Yet photons manifest as strings. These strings aren't the lifeless strings we know in our macroworld—for instance, the strings used in music instruments. A musician uses his energy to strike a string on a guitar to sound it. That musician's energy is transformed to the strings, to perform the work necessary to sound them. These strings have no life of their own, and never sound themselves.

An elementary string, as the elemental particle in our universe, is an object with intrinsic energy to perform work. This means that strings have their own energy that lets them move in different ways and therefore form different shapes, all on their own. Movement in these strings is inherent, and their basic movement should consist of spinning along their long axes. Other motions can be combined with or develop from the spinning, producing many different effects. I've already described some of them.

Motion is displacement, and life is the existence of the ability to vary that motion. Hence, life already exists in elemental strings. This properly is intrinsic, acting as an elementary eternal dynamo, able to create dynamic shapes that form our world and sustain our existence.

If each string spins at the same angular speed, they must contain some sort of basic clock that controls this speed. This establishes the length of a string's life, as a string clock will have run steadily from the first moment of its existence. This string clock does not tick; it spins. The standard spin is the time required for one spin at the beginning of its existence. This is the most basic definition of time. Any additional activity occurs after a certain number of those standard spins, thus after some amount of string time. The existence of strings and their activities could register on a scale reading in numbers of standard spins. Thus, strings have their own scale for recording their life activities. Every scale has its beginning; our string scale also has an origin time, a beginning when strings came to life. This determines the longest timeline scale, in terms of age, that could exist in the universe.

Let's look at the possible lifetime of a string, now a photon, that has been emitted by an electron. According to our understanding of nature, the string's lifetime began when it was shaped and started spinning. We know that in some cases, strings can be packed into a very small volume indeed, and that such objects have extreme density. Natural black holes have such densities. So do artificial compact objects created by baryon collisions in the LHC at CERN; indeed, the LHC recently produced matter of the highest density ever recorded by science. A string in a primordial black hole, created when the universe came into being, has already spun there many, many times. Strings began to be emitted from it as gravitons as entropy began.

Eventually, the string entered into a state where was not subjected to such extreme pressure, allowing it to increase its black hole volume and begin to move, though in restricted paths. The emergence of this movement gave dynamic properties to strings in that ancient world, and these dynamic properties have been retained since then. The primordial particle of ordinary matter was the neutron, and therefore the first mass appeared in neutronium. Then, our string found itself on the surface of a neutron, with one end fixed and the other free to move. Consequently, the free end began to wave out from its axis of rotation. Growing displacements from the axis transformed the string into a three-dimensional object—a transverse vibrating string. The "waving" displacement grew until it stabilized as a vortex, resulting in the first negative electric charge on the surface of that neutron.

If a neutron isn't bound to other baryons via the strong nuclear force (as already described), it decays into a positively charged proton and a negatively charged electron, plus a neutrino (and free strings). When this happens, our string ends up on the surface of the electron. After many more spins on the string clock, our string might be emitted from the electron so that it can travel freely in space. The free string becomes a photon. Besides the string, the electron also lost the energy that held the string in place: the first elemental entropy related to photons from electrons.

The string still exists, but in a new form as it travels toward the edge of the universe. The string is still live, and time still exists for it, measured by its "string clock."

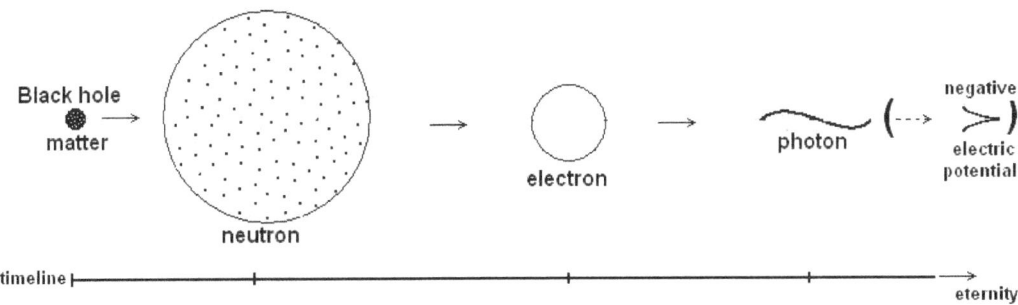

Einstein notes that time is related to the speed of light, and that this applies to photons. But it is obvious that *time existed before the existence of light*. In accordance with the example described above, the string was first bound into a black hole, then existed in a baryonic state for some period, then in an electron, and lastly as a free-traveling photon.

**Light Scaled by the String Clock**

The inner motion of the photon is displaced as it moves through space. This movement should still be evaluated in relation to the string clock—that is, to the string time. Now, speed is defined by the distance an object travels during an interval of time. Therefore, the speed of light is the distance traveled by photons divided by that interval. According to String Theory, there must exist a standard interval of time, $t_s$, the interval during which the string makes a standard spin. According to this time standard, the speed of light, **c,** in a vacuum is the distance, **a,** that light travels during a constant period of time, $t_s$.

In spite of this fundamental truth, Einstein and his followers claim that the speed of light in a vacuum is constant, and thus independent of the motion of the light source and light detector; and therefore, time itself can change—it can shrink or stretch.

This is not in accordance with natural logic (physics), since it claims that **the constant speed of light is the distance traveled during changeable time**. Or else a constant time does not exist, since it can change, and is variable depending on the speed of a light emitter or receiver.

Otherwise, the distance **a** is the wavelength of the photon, and therefore the wavelength of light, and the spin interval $t_s$ is the period of one circle. Consequently, the formula for the speed of light is: $c = a/t_s$. Science uses this formula specially for electromagnetic waves (perhaps for some, the familiar formula is c=lf, where l is wavelength and f is frequency, which is reciprocal to the period of a cycle).

Again, speed is the distance traveled by an object during a particular interval. Here, the object is the photon, and it travels a distance **s** that is equal to the **y**-multiple of its constant time, $t_s$, times the constant speed **c** ($s = yt_sc$). The length should never change, since length is always a multiple of the standard of length. One meter is 1/10,000,000th of the meridian through Paris between the North Pole and the Equator; and so this is the standard for length according to the International System of Units (SI), dating from 1793. This length does not depend on any type of motion. Prototypes of meters were also created for other countries, and I'm sure that these prototypes were no shorter as they traveled to those countries (due to the proposed contraction of length brought on by speed).

## The Second by the String Clock

Our SI unit of time is the second. The second is the interval it takes for light to travel 299,792,458 meters in a vacuum. If the speed of light is replaced for **a** meters when the standard string makes one spin, then the second is 299,792,458/**a** standards in string time, thus so many $t_s$. Clearly, **the second is just a multiple of standard string time**. And since the speed that strings spin (and thus their frequency) is independent of any speed of propagation through space, the length of the second cannot vary while the string propagates through space.

The problem here is to keep this prototype (the spinning string) in mind, since it is beyond our ability to observe so small an object that is rotating so fast. Human-scale prototypes of clocks work on physical processes, and these result from physical conditions. This is why we need to add to that standard clock the referenced conditions in which it should be working, and in which it will slow or speed up. Thus, we must reckon with the difficulties in observing the standard time, the standard "ticking," and even the possibility that enemies of science might create conspiracy theories against nature.

## Measuring the Existence of Life

Time serves to scale our existence. In the previous discussions, we looked at the activities of strings during their existence. Quantum-level life is based on the perpetual activity of strings, which have inner motion that may change their outward forms.

In going further into microworld, there are atoms and molecules that make up ordinary matter, and gravitons and photons (bosons) that serve as carriers of forces, acting on distant objects, as well as strings creating the static effects of electrical and magnetic forces. Atoms and molecules contain live strings, although they are not live units in the macroworld sense. This is because the primitive life of strings in enclosed in the volumes of subatomic particles, so as to be conserved. Still, strings living in these "containers" give them properties (mass, electricity, and magnetism), and carriers of force apply their force on the whole "container." For example, gravitons can make these "containers" heavier, since they give them weight.

Subatomic particles ("containers") do not have life as such in the macroworld, since the interactions among them work in accordance with laws created by the live strings (mass inside particles, electricity and magnetism on their surfaces, forces carried by bosons, and so on). Thus, the eternal life of strings creates the rules for our macroworld. Further, nature has a physical existence measured by string time.

Consider the form of life we see in the cells of plants. We speak of living plants and dead plants. A living plant is one that, as an object of our macroworld, is able to undergo natural processes, and thus create outward variations: for instance, to take nourishment and water from the roots into the plant, helping it grow, and to take further nourishment from photosynthesis, which creates cellulose and sugar to allow the plant to grow farther, into forms like wood, leaves, flowers, seeds, etc. A plant's life activity may also reproduce new plants. Material life is not eternal; it therefore ends according an internal schedule or can be destroyed, and so the plants die.

Higher forms of life have more variants, since they are not rooted to one location. Their inner needs to obtain energy for life are processed into motion, activities used to acquire food and other necessities. Still, the existence of their lives in the material world is not eternal, and that is why their lives complete in a finite time period. Therefore, the general consensus is that living beings live a temporal life.

Human beings have eternal activities within us, however, since we are also spiritual beings. These eternal activities provide our religious and spiritual lives. Here there could be hidden the principle of life, because we first come to eternal existence in the macroworld. An eternal principle of life would manage the activities of dead particles or "containers," causing them to live longer and to have an easy life.

Just as we do not directly see the life activities of strings inside particles (containers) and can still exploit them (e.g. via their mass, electricity, magnetism, etc.), we do not directly see the activities of the human spirit; but they register their footprint as live cells. Just as the mass of one subatomic particle contributes to a mass of a complex chemical molecule, many cells in a creature contribute to a higher form of life, growing in complexity from germs to plants to primitive animals, and so on.

Since strings having an eternal life created the physical laws of the material world, the eternal spirit created *spiritual* laws for the life that operates in human beings. Every normal human has his or her inner spiritual restriction against deeds that would hurt life (a conscience). These laws work on the principle of reciprocal action: Do not destroy or hurt life, because the same can happen to you. They exist also in written form: life for life, hand for hand, eye for eye, bruise for bruise, and so on. These spiritual laws in us are what tell us our life has an eternal source.

When an assembly of strings is broken (in subatomic particles), the strings don't cease to exist (this can be seen in the products of shattered particles in accelerators). This is also how it should be with life in the macroworld. When our flesh is broken or otherwise stops producing life activities in the material world, our spiritual substance lives on. Thus, in higher forms of life, time does not stop—it continues to run, providing existence of a further life, the eternal life. This means that if we were to measure the existence of life in time, it would continue until eternity.

For instance, my life started at some point. Photons were present at my birth. Since they bear string life, they registered my birth. Since they travel in space, they could carry this information into space at the speed of light. When photons bearing this information were 299,792,458 m far from me, or 299,792,458 m ahead of photons that were then passing me, there was a portion of their path 299,792,458 m long, and it was counted as one period of my life. The first second of my life is that first period of my life. Thus, the scale for my life has had its beginning, and runs into eternity – into the far universe.

When I die, the photons that witnessed my birth may be, say, 80 light years distant from the Earth. My death occurs in the material world, but the photons that witnessed and recorded my birth will continue on their way into infinity (eternity); similarly, my life will continue in the direction of eternity.

Eternal time exists independently for strings, whether they exist as part of a neutron, in an electron, a photon, or a negative electric charge. Even if the physical world should end, time as used by the physical world does not end. Since life in the material world has a source in the spiritual world, then human time also has its source in the spiritual world. Since string time is the basis and basic time unit for the physical world, where a second is a multiple of the string standard, spiritual time should exist similarly. A second of that could be $(24 \times 360 \times 1000) = 8,640,000$ our seconds, or 144,000 minutes, or $24 \times 100$ hours.

This leads to the conclusion that when a life starts, time begins to run for it. This time does not end with the death of the material flesh, but continues into eternity. Thus, no speed (even the speed of light) can destroy time that displays the existence of lives and life activities performed during that time.

**Delayed Photons**

Photons—including light—usually propagate through a vacuum at the speed **c**, which equals 299,792,458 meters per second. However, the speed of light is sometimes lower than it is in a vacuum. We know for a fact that light propagates through some transparent materials more slowly than it does through a vacuum. Hence, photons must interact with matter to cause some delay in their propagation, as related to the normal speed of **c**. The speed deficit seems to result from a loss of some speed in relation to the photon's regular progress, but the physical truth is that the instantaneous speed of light in transparent materials never truly falls below **c**. The difference is caused by "breaks" in the path of the photons, during which they interact with electrons in the material. The electrons give the photons a "ride" for an orbit; and since they spend some time spinning around atoms with their electron partners instead of propagating in a straight line, this seems to slow them down.

To be clear, the photons never slow their movement; they just change trajectories. When a photon collides with an electron, it's captured briefly, and its straight-line movement is interrupted as it begins a circular movement as part of the orbiting electron—much as a roundhouse might change a train's straight path into a circular one, then send it back to the straight route. When considering this fact, we also need to consider the direction of movement and, therefore, velocity. If the photon never slows down, then the electrons it interacts with should also orbit at **c**—the same as the velocity of light, since the particle and the string interact at the same velocity. However, the electron is also spinning; therefore, the velocity of the surface strings riding on the electron should be added to the orbital velocity, which together would be equal to **c**, the speed of the photon (light). When interacting, the photon and the electron orbit the atomic nucleus as a couplet. This couplet is excited energetically, and consequently, when they reach the velocity that the photon had before interacting, the photon is emitted and continues on its straight-line path.

From an outside perspective, photons propagate slower through a transparent or translucent medium than they do through a vacuum. Hence, we register photon velocities lower than **c**, but never higher than **c**. This aspect of light affects our world significantly, where we must always assume some delay in any activities connected to light. Since our life activities are connected to light and even scaled by light, physics must reckon with this fact. Otherwise, physics has problems in understanding our world—that, two types of physics coexist.

Life activity also registers information brought by the photons, recorded by their interaction with matter and so on until we acquire its output as perception. The output of information always comes after the non-interfering photon (traveling early compared to the photon that entered into interactions) is far from us, and new photons enter into this process of being perceived. Thus, *our life reality is always delayed in relation to the reality of light.* This delay was caused by the interference of the photons with the electrons of our body; thus, additional time is needed to reach the processing place (the mind), where the processing of information from our senses also takes a finite time. Therefore, *our perception of nature is inadequate to the reality.* It's like how we never find the fish underwater where we see it. In such a case, no one claims that space itself is bent; no, the incoming light from the fish appears to be bent due to light's differing velocities through water and air. The **water** refracts or bends the light.

This also takes place in other situations, but those situations are often used, in the new physics, to perpetuate the fiction of bent or otherwise distorted space.

The delay of the output to the received input of light can be compared to ordering pizza and consuming it. The input occurs when the customer orders a pizza. It takes time to get the pizza, so the consumption is delayed. This is because the order must be completed, a cook must make the pizza in a kitchen, and then the prepared pizza must be delivered to the customer. However, if pizza were made during delivery, the delay would be shortened. It's possible to decrease the time that passes between ordering pizza and consuming it. But does time itself run quicker in this example? No, the amount of time that has passed is just shortened due to manufacturing the pizza in the moving car. The delay between purchase and consumption can be decreased, but the real time involved never changes.

**Dilation of Perception**

Einstein failed to consider the delay inherent in perception in his theories; and therefore, different observations in relation to speed led him to create his new physics for extremely high speeds. However, if we were to sit in a car moving at the speed of light parallel to the direction in which the photon delivering the information travels, then the processing period would not play a role, because the output of information is on the same spot where delivery started. Therefore, this postponed perception of nature ends at the speed of light, and not time. A perception of events in time is not equal to the real time that runs independently of us, because *that* is rooted in string time. **Since we sense our world in a way that lags behind real time, we need to use speed to eliminate this lag.**

Then, of course, different speeds may create new disagreements that give the impression of dependence on speed.

The example: Our life activities are activities performed in real time. But they run at very slow speeds, and therefore there is very little if any delay to force us to think that we're lagging behind these real occurrences in nature. Let's say our perception delay is 10 nanoseconds ($10 \times 10^{-9}$ s), so we're 10 ns behind the reality. Still, at a speed of 100 km/hour, 10 ns gives a distance of 30 nm (0.03 micrometers), so we're still not very far behind nature. But if nature, meanwhile, speeded up from 100 km/hour to the speed 1/3 c, then Newtonian physics would not produce exact results, because our distance delay has increased from 30 nm to a distance of 1 m. We see the distance shortened about 1 m in relation to our expectations. Therefore, a real object moving by a high speed might seem to us to be shorter. But this apparent shortening of length is just a fiction of perception.

Similarly, consider the speed of a clock on a very fast rocket. Since the speed of the rocket is known, it seems to observers that time has slowed as the rocket's speed increases. Therefore, we may think that time in objects moving at high speed have changed in relation to our time on the Earth, and that time runs slower in the rocket's reference frame – resulting in a *dilation of time*. But in reality, time never increases or decreases. The fiction that a cosmonaut in a rocket lives longer, due to a prolonged standard of time, does not meet the criteria of string time and eternal time.

The correct assumption is that we're seeing **a delay in the perception of nature due to the limitation of our senses. Actual time dilation does not exist, any more than dilation of length does.**

**The Speed of Moving Strings**

Observing life leads us to the conclusion that processes run in time. Since strings are the smallest elementary parts of matter and energy, the elementary life in strings causes the elementary processes to adjust themselves so that existing needs will be satisfied. That means that there are decisions made for incoming information, and these decisions are realized. The chief need of live strings is to always be in motion, never to stop. Since string life is scaled by the string clock, then all its activities run in real time. For instance, free strings should show these activities:

1.  The primary movement is the spinning of all points, as they would spin an axis. Time is related to the frequency of the spins.

2.  The next basic movement is changing the densities of the points on a line (stretching and shrinking, as a spring moves). Thus, the string oscillates in a one-dimensional, longitudinal vibration (back and forth). In relation to time, there is a frequency of vibration.

3.  A point of a string also moves in a direction perpendicular to a line when the string spins, and this it vibrates transversely. Consequently, there is a frequency of transverse oscillation as well.

4.  As it spins, the string may move at one end so rapidly that it forms a circle as it rotates. The end of the string circles around an imagined point called the point of rotation. The string may rotate like a wheel. The string has an angular speed, and at a large scale this is the frequency of rotation.

A string's ability to move means that physically, there is energy in the string. Energy suggests the ability to do work, and so to produce force. I have already described the elementary forces of strings: the electrical force and the magnetic force. When they act, these forces create impulses during time (impulse = force x time).

The abovementioned inner activities of strings are independent of the actual speed of propagating strings, since these represent their life energy. This means that if free strings create the lines of an electrical field around electrical potentials—say, the ends of an electric battery—then if this battery (in this frame) were placed in a rocket moving at the speed of light, the speed of the rocket in space will *not* change the activities of the string life that create lines of the electrical field. The electrical field of the battery frame is the same in the rocket's frame as in the Earth frame. The change of the speed of an object's frame never changes any life activities of the strings inside this object's frame. Hence, the changed speed of a frame in nature never prolongs or shortens the string life inside this frame. That is, the **time dilation effect does not exist for the elementary life or activities of strings.**

Besides the elementary life inside strings, there exists also the possibility of the strings being more rapidly displaced through space. Thus, quanta of energy have a speed. They travel in space at the constant speed of **c**, 299,792,458 meters per second.

A graviton is a quantum of energy "living" in the second kind of kinetic shape (probably in the first, also) and moving at **c**. Hence, the inner life of the graviton is vibrating at string points, just as a mass connected to a vibrating spring does. The average speed of the ends of the string may be equal to **c**. Since there are rapidly changing directions of movement, there are positions when a string's endpoints stop and switch direction, and therefore the endpoints should have their maximal speeds at the equilibrium points of their movements—in the middle of their routes. Thus, the maximal speed there should, for a moment, be higher than **c** in order for the **average speed** to be **c**.

This vibration is not affected by moving gravitons in space, and therefore the (inner) life of gravitons does not change during their propagation through space; the life of the graviton is in the one frame (graviton), and this frame moves at the speed **c** in another frame—space. In the same way, blood flows at the same speed in a passenger's veins independently, whether he is in a moving passenger train or stays on a trackside platform.

We may calculate both speeds—the speed of a propagating quantum of energy, and the speed inside this string. Putting them together, the resulting velocity for the end points of our string is sometimes higher than **c**. Therefore, **c** must not be the maximal possible speed for gravitons in String Theory.

The string, vibrating like a spring in its frame of reference, can also have a restricted degree of movement (displacement of a whole frame) when it is "locked" into the volume of a subatomic particle. If their restricted movement is random, like Brownian movement, then it is a zigzag motion. If the movement results from the speed constant **c** for strings propagating in space, then there must be acceleration and deceleration that, together, cause the average speed to be **c**. Hence, there must be instantaneous movement **preceding** the speed **c**. Adding together the inner motion of the string (where there exists a higher speed than **c**) and its movement in the volume of subatomic particles (where there also exists a higher speed than **c**), we must achieve a higher instantaneous speed than the speed of light. This is the speed at gravitons.

A quantum of energy "living" in the third kinetic shape is a two- or three-dimensional string. When it propagates through space, it is a photon. We can picture this photon as a two- or three-dimensional sinusoid wave as it propagates through space. Since this waveform is very well known, the conclusion is that **c** does not change the activities of a string vibrating transversely.

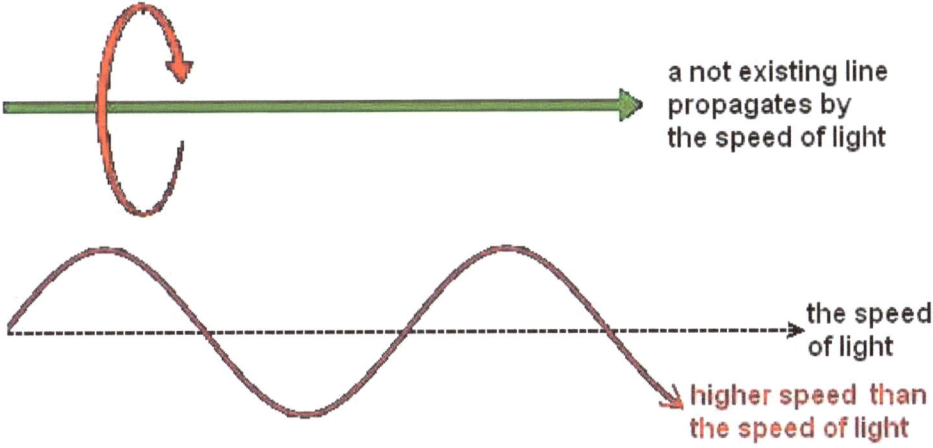

a not existing line propagates by the speed of light

the speed of light

higher speed than the speed of light

Yet we may calculate both speeds—the speed of a propagating quantum of energy, and its interior speed... its life. Adding them together, we find that the resulting velocity for any point of our string to be higher than **c**. Therefore, **c** must *not* be the maximal possible speed in String Theory. Only in relation to quantum theory may we say that quanta of energy move at the speed **c**, since the inner life of a string occurs inside the quantum frame.

But let's look at the fourth dynamic shape. The inner life of a string in this configuration is characterized by circling. Each point of this string may move in a circle at a velocity equal to **c**. A circled string does not propagate as a closed circle (unless they are bound into a neutrino) because a circle breaks; it becomes open, and thus sinusoid (see the image of the modeled movements of the photon).

The real trajectory here is not a straight line, but an undulation—a wave. Two routes are combined here. One is a straight line, and the second is a circle. If the straight route has the velocity **c,** then the undulating route is the combined straight and tangential velocities, and therefore has a higher speed than **c.**

Suppose we start with two cars on a starting line; one is set to take a straight-line route, the second to follow an undulatory route the same distance. If they end up at the finish line at the same time, then obviously the one on the undulatory route must have traveled faster than the one on the straight-line route.

Einsteinian physics does not allow for the existence of a higher speed than **c.** But we can easily see that this limit is a false one, so modern physicists must delete the postulate that **c** is the highest speed allowed in nature from their theoretical constructs if they want to make any progress in physics and String Theory. Otherwise, they deny that light is a string that propagates as a wave, which we *know* to be true.

Many scientists appear to be schizophrenic when they link quantum and String Theory. The source of today's schizophrenia in physics is their non-uniform theory of light. They picture light as both a particle and a wave. To lock it into unvarying definition, they add mathematical definition to the quantum. That way, they substitute the physical bodies of the concrete world (the string) with mathematical abstracts. But this makes things worse, since they are backing away from physics into mathematics. Because they divide physics this way, we're still unsure whether it's the mathematical abstraction of nature (quantum mechanics) or the physical reality of nature (Newtonian mechanics) that matters the most. All these differences can be accounted for, giving concrete forms to physical particles that move and interfere with each other. Therefore, particles and forces are not the result of the standard mathematical model of particles and forces (the Standard Model), but the physical results by which strings behave.

A string, by then a quantum of energy, has its own inner life. A wave is just a trajectory; thus, light does not propagate in a straight line per quantum theory, but as a wave, per string theory. The photon is not a wave as such; it is just a portion of wave, so the string has its length, called its wavelength. As this string propagates, since it has its inner life, it moves like snake on a waving or undulatory route. The string is an object having inner life. Light *is* the string, having its inner life in the third and the fourth kinematic forms and propagating through space on a trajectory pictured as a wave. As described, light (the string) has a higher speed than **c.**

I conclude that for String Theory, it is unacceptable for the speed limit of light to be **c.** It must be higher.

## The Light Field

In our solar system and world, light comes almost entirely from the sun, and therefore we register the speed of light as equal to the speed of photons emitted from the sun. But light, and therefore photons, are strings traveling as such a way that they appear to be waves—something like a train riding on a railway with a lot of switchbacks. The trajectories of propagating photons are waving lines; this is caused by their frequency of rotation as they travel.

Since the solar system is full of photons and they are of the stable intensity, then the same trajectories of photons should exist. These trajectories would create the geometric structure of this **light field**. These trajectories are not smooth, straight routes;

they're wavy (undulatory) due to the nature of the photons. The density of these trajectories increases in the direction toward the sun, and decreases away from the sun. The speed of light propagating on these trajectories is constant at **c**. It cannot increase or decrease, since the fabric of these routes does not allow it. To illustrate this, I'll use the example of a corkscrew twisting inside a cork.

The inner life of the photon is realized so that the string rotates around a virtual axis illustrating its direction of propagation. Just as a subatomic string twists in space, so does a corkscrew twist into a cork. If there are many corkscrews, then they may make many sine tubes (solenoid holes) in the cork. A screw that goes in after them might use existing holes, propagating through the cork more easily.

The inner life of a photon imparts a stable spin to the string—in other words, the same constant, angular speed (frequency). Now, if all the same metaphoric corkscrews have the same angular speed, then while twisting into the cork, they propagate at the same speed inside the cork. Again, when other screws travel through the cork, they may use existing sine tubes. If they have the same angular speed as the screws before them, they must propagate at the same speed as those screws. Another aspect of the "fabric" created in the cork is that if other screws having the same angular speed want to travel from the other side of the cork, they can, but must propagate at the same speed as those that created the holes.

This follows from the second postulate of Einstein's theory of special relativity, the Principle of Invariant Light Speed. For our light field, this means light propagates at the same speed whether it travels toward you or away from you. Reflected light travels at the same speed, whether it reflects from a mirror or the moon. Many experiments have demonstrated this, and Einstein used some of their results (for example, the Michelson-Morley experiment) when developing his theory. Einstein could not physically explain this invariant light speed, so he simply accepted what he could not understand. Then he took advantage of this mystery, forming the principles of the new physics sometimes called Einsteinian physics.

Thus, science stepped down from a physical understanding of nature, onto the route of mysterious physics. Consequently, just as one mystery was conceived, others then conceived more mysteries, and so on, until today there exist a forest of mysteries in theoretical physics. They include, for instance, changing time; changing space; the mysterious theory of gravitation in the general theory of relativity; bending space; expanding space; other universes; other worlds around us (the Many Worlds theorem); particles from other worlds giving certain properties (like mass) to objects in ours; the idea that particles of matter do not really exist, i.e., that they are just higher concentrations of energy; that space consists of at least eleven dimensions; and so on. To uproot these artificial mysteries, we must explain the initial mystery that caused this bewilderment in physics in the first place.

Since the speed of light, or $c$ (299,792,458 meters per second) has the sun (a star) as its source, this speed is an intrinsic property for all light fields created by stars in our universe. Therefore, the physical rule for light is that light travels at the speed $c$ in all its frames, in all light fields.

## Alien Photons Entering our Light Field

Besides the photons we receive from the sun, other photons come to us from distant stars. Thanks to them, we can see deep into the universe. These photons have the same frequencies and wavelengths as the sun's photons. This is because the same fusion of simple elements into heavier ones occurs in each star, beginning with the fusion of protons, the nuclei of hydrogen, into helium. If a distant star doesn't change its distance from our sun and, thus, from our sun's light field, then nothing changes the speed of its photons. But as we know, such a situation is rare; stars are in constant motion. They revolve, and rotate around the center of the galaxy. Thus, they may approach closer to the sun, or travel farther away. If so, what should happen to photons coming into our solar system from outside?

I suggest that photons entering the solar system from other stars must adjust themselves to the existing traffic of the photons in our solar system's light field in order to propagate through our light field. Our light field is filled with the sun's photons, laying down their trajectories through space along constantly reused routes. These routes are like railways issuing from the surface of the sun, and continuing in all directions. In free space among these "railways" there are created wave channels with conical volumes; they're narrow at the sun, but widen as they proceed away from the sun.

Suppose a photon from another star at a stable distance from the sun enters our light field. Its wave is moving at parallel to some of the approaching routes (trajectories) of the photons from the sun. Alien photons may avoid collisions with local photons by the way they propagate in the free space of wave channels, having the shape of undulating tubes. That is, alien photons use such "tubes" having the same wavelength (frequency between crests). Since these routes are parallel (the sun shines in all directions) then they all have the same speed of travel. Thus, photons from both stars (the distant star and our sun) have the same speed, although in opposite directions. I conclude that alien photons entering our solar system would enter and pass through **free tubes of the light field of the sun**. That is, the alien photons are forced to take existing routes through our light field.

But suppose that a photon arrives from a star that's moving toward our sun. The photon enters the light field in relation to our frame of reference at the combined speed of the other star in relation to our sun ($v_s$), and the speed of light ($c$). These alien photons having higher speed ($c + v_s$) propagate face-to-face to the sun's photons. But these alien photons will not strike our photons, since they can glide alongside them. Thus, alien photons enter free tubes in the local light field at a faster speed than our photons. As they continue on, however, the shape of the trajectory tubes in our light field forces them to slow down as they strike the "sides" of the tubes. As the trajectories of photons of all stars, and so also the free tubes in the light fields of all stars, have the same wavelengths, the alien photons will also have the same wavelengths. If the other star's photon trajectory and the sun's photon trajectory are on a line, then the trajectory of the sun's photon does not change, but the trajectory of the alien star's photon has an extra component to the velocity ($v_s$). Placing the alien photon's trajectory inside a free tube in the sun's light field will quickly eliminate this extra speed.

In doing so, the alien photon that enters our light field may appear to change its wavelength, but I don't believe this is true, since this would change the identity of the alien photon or may even to destroy it (by liquidating its wavelength and so on). In the example of the corkscrew, the alien photon is a "corkscrew" propagating though a cork (our light field) that already contains holes created by other corkscrews. This alien corkscrew is coming from the opposite direction. Since it has a higher speed than our photons, once it enters a sine-tube (solenoid), it pushes against the sides of the solenoid, which creates friction slowing down its propagation. The friction ceases when the alien photon achieves the same speed through the sine tube as the photons emerging from the sun in the opposite direction.

The alien photon enters the narrowed channels, and twists around in them. These twists bit by bit correct the alien photon's very small advantage in a speed, so that the alien photon does not disturb the photon traffic in the solar system. The same occurs if a fast-moving car enters a highway where the other cars are traveling slower; it must lower its speed to the speed of the other cars, especially when those cars are moving in a group.

Therefore, we register photons coming from other stars at a velocity that is independent of the state of motion of the emitting bodies. The same effect is used in part with Cherenkov neutrino detectors, which register photons as they adjust to the speed of light in the medium observed.

On the other hand, when a distant star moves away from our sun, then the light field in our solar system forces the photons of that star to speed up a little to adjust their speed to the new frame of reference within the sun's light field of the sun. This is like a slow car entering the highway having to speed up in order to merge safely.

## Interaction Between Two Light Fields

The cases described above are for individual photons from other stars entering the light field of the sun (our star). Since the universe is huge, there are many interacting light fields. Their interactions would be especially noticeable in multiple star systems in which the stars are close to and orbit around each other.

Each star possesses its own light field. When alien photons enter the light field of the solar system, they are forced to adopt the trajectory and speed of our light field if their star is not perfectly stable in relationship to ours. When an alien photon has a higher speed in relation to the speed traffic in our light field, it pushes into our light field. According the Newton's Third Law, there also exists a reciprocal pushing, so the light field pushes against the alien photon until it slows down to the local light speed. Because the alien photons from a particular star are very few in number compared to the number of local photons, their pushes against our light field are negligible and easily overcome. But what if a number of photons were to arrive from that source at a rate similar to the number of photons emitted by our sun? In that case, the "push" from the alien photons might replace or deform the light field of the sun. This has never happened to our sun's light field, since nearest star is 4.37 light years far from the Sun. Yet this is a binary star, Alpha Centauri A and Alpha Centauri B. There, the fabric of their light fields could be quite complicated, especially since the two stars are similar in size.

The light fields of such stars must merge and force **photons of both stars to travel on the newly created trajectories**. The observed result, then, should be that there are no speeding or slowing photons when one star (let's say Alpha Centauri A) is moving toward us and the other (Alpha Centauri B) away from us.

Therefore, since both stars are in dynamic movement, the dynamic movement of both light fields creates an interference field that is dynamic in relation to the density of photonic routes in the light field.

Something similar could also happen on Earth, when other sources of light shine brightly. They might create new local light fields that last for a short time.

The above-described light fields are not very dense, however, since the density of photons is not very high when compared to, say, quasars (quasi-stellar objects). Quasars are extremely luminous, so their light fields are so dense that any alien photons must certainly respect the traffic there. In our solar system, although the sun is quite active, the density of photons (luminance) falls by the square of the distance from the sun. This means the light field at Earth's orbit is so sparse that photon traffic at the Earth can exist essentially without any speed and direction restrictions.

## Bending Light

Einstein determined that massive objects can cause a distortion in local spacetime, which is felt as gravity. One conclusion of Einstein's general theory of relativity is that gravity can affect light, causing it to bend; this suggests that light can be bent entirely around a massive object, and therefore may act as a lens for the things that lay behind that object. They call this phenomenon "gravitational lensing." Supposedly, this allows us to see an object (a quasar, for example), even when a massive object like a galaxy blocks our direct view to the quasar.

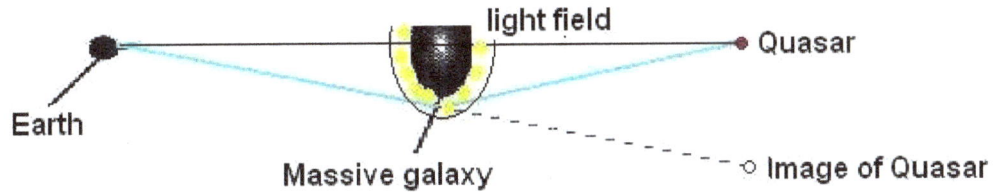

Such lensing has been observed through powerful telescopes, but I would argue that it is *not* caused by gravitation. Gravitons are one-dimensional, since they vibrate longitudinally (back-and-forth), as a spring does. Photons move in transverse (side-to-side) fashion as they propagate through space, so they are at least two-dimensional, and more likely three-dimensional. Two-dimensional photons vibrate in a perpendicular motion in relation to the direction of propagation. Thus, although the same inner life is present, gravitons rarely interfere with such photons due to their perpendicular motion. Theoretically, if a transverse string is transferred from three dimensions into one dimension, then it becomes the same as a graviton, and then they may interfere. For this reason, longitudinal strings may theoretically interfere with each other in some situations. However, I will not consider this in my discussion, in order to make it clear that the concept of the gravitational bending of light is *not* true physical theory.

I assert that "gravitational" lensing is actually due to the well known optical phenomenon of refraction. Astronomers claim that a galaxy produces gravitons, via the many bodies it contains, and that therefore there is not a homogenous gravitational field in a galaxy. A maximum amount of gravitons can produce the largest possible black hole in nature at the center of a galaxy. The gravitational field of the black hole holds the galaxy's bodies in place around itself. There are many stars in a galaxy, and all are producers of photons (not just gravitons). Each producer of a large density of photons creates a light field around itself. Therefore, a light field exists in conjunction with the gravitational field in each galaxy.

## Bending of Light in Other Light Fields

As I have noted, an alien photon propagating through a new light field must respect the restrictions of this field. The first restriction causes a high density of trajectories (routes) of photons in the light field—a high traffic. The highest traffic will be close to the shining object. As pictured in our illustration, the quasar's light enters the denser traffic created by the massive galaxy's light field.

Quasars are moving away from us as the universe expands, and therefore, as previously established, the quasar's photons propagating through the galaxy's light field must increase their speed in the galaxy's field. The changing speed of light causes the light to bend according to the principle of refraction (assuming there is an angle of refraction to the angle of incidence). Speeding up the lightwave causes the change in direction. This is how the galaxy lenses the incoming light—by changing its speed. There is no need to invoke a gravitational interaction with the light itself.

## Bending of Light Due to Propagation in a Denser Medium

The physics of the laws of refraction determine how light entering an optically denser medium will bend. Optical lens work using these principles.

Some scholars claim that the bending of light around massive objects is experimental evidence of Einstein's General Theory of Relativity. According to them, light should be deflected when it passes close to a massive body. They first observed this during a total eclipse of the Sun in 1919, and declared it the first verification of Einstein's general theory of relativity.

But the light they observed was not propagating through a vacuum; it was passing through a thin but extant medium extending above the surface of the Sun known as **the corona**. This corona extends for millions of kilometers into space. The corona behaves like a very thin gas. Therefore, light propagating through the Sun's corona **must slow down** due to the refraction index of the corona. During an eclipse, observers on Earth do not see the light rays emerging directly from the Sun, because they are blocked by the Moon. What we see are rays emerging from the sides of the Sun, which come to us at a slight angle. Since the corona is an optically denser medium than the vacuum between the corona and the Earth, these rays must slow down slightly while propagating through the corona. According to the laws of refraction (Snell's Law), they must therefore change direction.

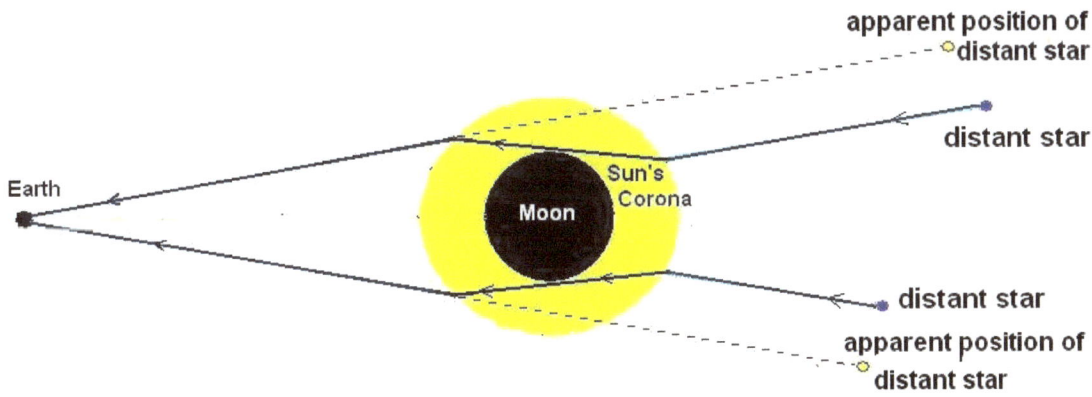

If a star is behind the edge of the Sun and its incident rays propagate through the sun's corona, then it is possible that the apparent position of the distant star will not agree with its real position. This is not due to gravitational lensing, but to optical lensing—the light passes through the optically denser medium of the corona and then into the vacuum beyond, where it moves slightly faster. The corona, according to physics, must change the direction of the starlight. That's why we see distant stars almost in line with the Sun in different locations than they should be in. The shift is very small—about one ten-thousandths of a degree.

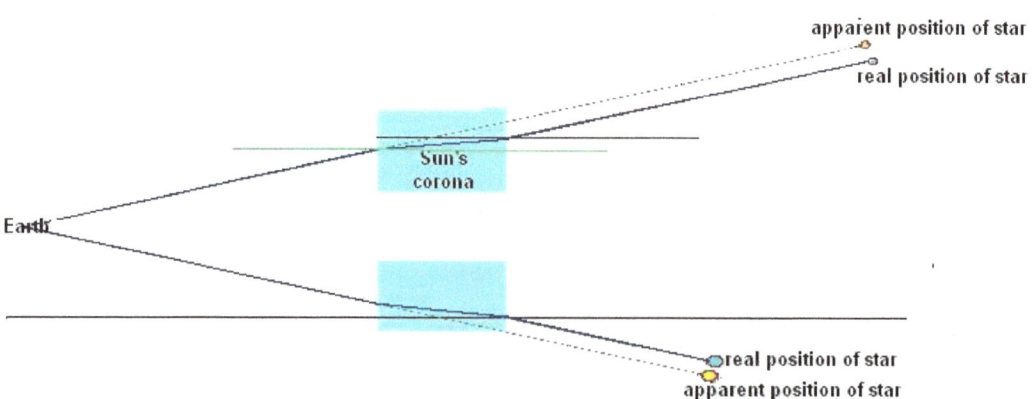

In a previous image, you saw the massive galaxy bending light coming from the quasar so that the observer on Earth could see the quasar, which would not normally be visible. This is natural physics, in accordance with the above-described refraction of light around the Sun. When light from the quasar must propagate through a medium that has a higher refraction index (e.g., is more optically dense), it must necessarily change its direction. The space within the galaxy is optically denser than the intergalactic vacuum, because it contains the light elements produced by stars. These light elements are mostly pushed by starlight into interstellar space.

We know the refraction index for hydrogen at normal atmospheric pressure. According to optical physics, when light enters a concentration of hydrogen from a vacuum, it slows down. The refraction index for hydrogen is 1.000132, which means that light propagates through hydrogen at the speed 0.999868 **c** on Earth. Yet light should also slow down in interstellar space, because there is a great deal of hydrogen there as well; indeed, hydrogen is the most abundant element in the universe. In any case, light propagating through or near a galaxy must bend in accordance with the laws of refraction.

When astronomers see, for instance, a quasar in the Pegasus constellation that lies about 8 billion light-years from Earth and sits behind a galaxy just 400 million light-years away, it is because the quasar's light is refracting through the denser medium in the galaxy. It is neither accurate nor necessary to explain the bending of light as the result of gravitational lensing.

Here's another thing to consider. Slowing light is perceived as redder than normal light; this is called the red shift. We can detect red shifts from distant celestial objects, which has led astronomers to conclude that those objects are all moving away from us. However, some astronomers have also found that this explanation does not fit well with the observations. They suggest that the red shifts may arise from some other process, and that therefore, the quasars aren't really so distant. This is called the "red shift controversy."

There's a very low density of hydrogen even in intergalactic space, due to it being pushed out of those galaxies by photons (dark energy). My theory of refraction in the universe explains the red shift controversy. Light is red-shifted not only because a shining body is moving away from us, but because light slows down in intergalactic gas clouds.

In conclusion, we can see objects hidden behind other massive celestial bodies because the light they emit is bent—not by gravitational lensing, but because those rays have to propagate through the optically denser media in and around the lensing bodies. This is not a gravitational lens effect, but an optical lens effect. We see this here on Earth, in which the images of objects are displaced because the light they emit or reflect goes through a medium (such as water or glass) with a different density than air. We exploit this natural property of light to make corrective lenses, camera lenses, magnifying lenses for microscopes and telescopes, and similar products.

No one with any sense believes that the bending of light at this scale is caused by the gravitational difference in the glass lenses. So why should gravitational lenses exist in the universe? Don't we also belong in the universe? If so, then we must also look to optical lensing to explain the bending of light in the universe.

**The Velocity of Light Propagating through a Medium in Motion**

Einsteinian physics requires light to have the same velocity in a moving optical medium as it does when that medium is at rest. In experiments, Einstein's followers have found that the velocity of light does not change in moving water, which confirms the second postulate of Special Relativity.

In an optical medium, the velocity of light is reduced to **c/n**, where **n** is the refractive index. Therefore, the velocity of light in a stream of water moving with velocity **u** should be **c/n+u.** Thus, adding the velocity of water to the velocity of light in water *should* produce a higher number than the velocity of light in immobile water. But experiments do not agree with this, and therefore the experimenters believe they've proven Einstein's theory.

Yet let's look at what real physics says. Photons interact with molecules of water, and thus with their electrons. The photons stop propagating on their normal straight-line trajectories while taking "rides" on electrons in orbit about one of the atoms in a water molecule, before letting go and continuing on their straight-lined trajectories. Thus, when light propagates on the straight-line trajectory, it propagates at the speed **c** during time $t_1$. When this straight-line travel is interrupted, there are added breaks, $t_{breaks}$, when the straight-line velocity is zero. Remember, velocity is the amount of distance covered over time. The distance covered is equal to the speed **c** times time, $t_1$. But there is also time required for all the breaks. Thus, the actual velocity is $ct_1/(t_1 + t_{breaks})$. The resulting velocity of light in water is less than **c,** due to the absorption and reemission of photons from water molecules ($t_{breaks}$). The discontinuity ($t_{breaks}$) depends on the number of molecules present to interrupt the progress of the photons. The number of molecules in water depends on the density of the water. Just as the density of the water does not depend on the velocity of the water, the velocity of light does not depend on the velocity of the water. Thus, the velocity of light in transparent materials is independent of the velocity of those transparent materials. Hence, these experiments do not prove Einstein's special relativity.

**Time at Higher Speeds**

Einsteinian geometricians claim that time is a physical unit that changes due to speed. Thus, time is not absolute; and therefore, a formula must exist that will allow us to determine the magnitude of time. Since we use clocks to measure time, they claim that a clock in motion slows down relative to an immobile clock. They've proven this by observation of muon decay in rest and in motion.

Decay is a process involving the death of a particle—but dying still belongs among the processes of life.

Most naturally occurring muons that reach the Earth are created by cosmic rays, which consist mostly of protons. When a cosmic ray proton impacts atomic nuclei in the upper atmosphere, of the Earth, pions are created. The pions rapidly decay to muons with a lifetime of $2.6 \times 10^{-8}$ s.

Protons travel at the speed of light; and, having mass, they also have linear momentum. Therefore, mouns also have a velocity near the speed of light, as do the pions produced from these high energy cosmic rays. They also continue in the same direction as the original proton.

A muon's lifetime in a laboratory is about 2.2 microseconds. This lifetime would allow a survival distance of only about 456 m in a vacuum. Natural muons come into being at a height about 16 km about the Earth's surface, so such a lifetime should not allow those muons to survive to reach the Earth. Yet about 10,000 muons reach every square meter of the Earth's surface per minute.

To explain this phenomenon, physicists theorize that the lifetime of those muons has changed, but only in our frame of reference; that is, time has dilated so that they live longer from our perspective, though not from their own. According to Einsteinian theory, these processes run in the moving muon at its own "clock rate," which is much slower than the laboratory clock rate.

Such a theory does not take into account the physical view of this decay process.

A muon is an unstable elementary particle with the negative electric charge, like that of an electron. A negatively charged particle cannot easily propagate through matter, because atoms are surrounded by electrons. Because these charges repel each other, the muon must push against the electrons in the air molecules to penetrate the air (the electrons repel each other). Hence, the muon faces constant barriers when propagating through the air.

A muon has a mass of 105.7 MeV/c$^2$, which is about 200 times that of the electron. A muon decays into an electron of the same charge as the muon and two neutrinos of different types. The resulting electron has 1/200 of the linear momentum of the muon, so 0.5 percent of that momentum goes into the electron, and around 50 percent each is carried by the two neutrinos. If muons have linear momentum in the direction of the surface of the Earth, then the momenta of the new decay particles are also directed toward the Earth. This process should run in such a direction that it meets the air at a very high speed (about 0.9997c). It is clear that objects moving at such a speed would experience enormous pressure from the direction in which they move. The extreme pressure means extreme force on the reactive surface of the muon, and therefore this force does not allow the muon to expand and decay. This is why it is physically impossible for the muon to decay while traveling at nearly the speed of light in a thick atmosphere.

A muon is not an ordinary particle, because it cannot stably exist in the universe under normal conditions. The muon is a temporary particle that soon decays into stable particles. In other words, the strings comprising the muon are not in a stable configuration. The stable configuration for strings vibrating longitudinally (so that the strings contribute mass) is a ball, similar to the way we register the shape of a drop of water created from molecules of water. But since muons are not stable, their strings are not collected in a spherical volume.

Since the muon decays into an electron and two neutrinos, from this viewpoint the muon is a transient "compound" of an electron combined with two neutrinos. The electron has strings vibrating longitudinally inside its volume, and funnel-shaped rotating strings on its surface. Since the muon is 200 times heavier than the electron, the funnel shaped strings (the surfaced on the electron) are relatively few in number. Besides the electron, the products of the muon are the electron antineutrino and the muon neutrino.

The neutrino has circled (looped) strings creating short tubes that cover a spherical surface. The physical property of looped strings rotating like wheels creates an elementary magnetic force. Still, these wheel strings have linear momentum if the neutrino propagates along a straight-line trajectory. Earlier, I depicted the dimensional image of a neutrino; if you refer back to it, you will notice where the problem is with the instability of a particle containing two neutrinos. The muon is unstable because of the wheel strings that prevail there.

Wheel strings stay together *only* if they rotate in opposite directions, due to the principle of magnetism. Therefore, the instability of the muon is apparently caused by the fact that the wheel strings rotate mostly in the same relative direction. This result supports the creation of a (muon) neutrino and its antineutrino. They differ only in the orientation of their spins. Neutrinos have left-handed spins, and antineutrinos have right-handed helicity. Hence, the instability of the muons is caused by impaired strings. Strings spinning in the same direction repulse each other. The assumption here is that a repulsive magnetic force is responsible for muon decay.

The repulsive magnetic force existing in the volume of the muon separates the neutrinos. The impulse of this disjoint lasts a few microseconds ($2.2 \times 10^{-6}$ seconds) in a laboratory. But this lifetime is for the frame of reference of the laboratory. The muon in the laboratory does not travel, and that thus the repulsive forces within the muon are not restricted. Restrictive forces come into play during movement. For example, as we pile on speed in our macroworld, we feel the pressure of the air on the surface facing in the direction of the movement. The magnetic repulsion normally pushing neutrinos apart is restricted, and therefore cannot push those neutrinos apart.

In the laboratory, the restrictive force depends on the Earth's atmospheric pressure. Atmospheric pressure at the surface is about $10^5$ Pa. The air drag (or viscosity) caused by the speed of movement does not occur in the laboratory frame. A muon created by cosmic rays has a speed of 0.99**c,** and therefore experiences an extreme air drag on the surfaces from which particles would otherwise be emitted. The pressure of the air drag is **0.5Crv$^2$**, where **C** is the drag coefficient (0.5 for a sphere), **r** is the density of the air (0.46 kg/m$^3$ at 11 km), and **v** is the speed of the muon, equal to 0.99**c.**

Thus the restricting pressure is $(0.5 \times 0.5 \times 0.46 \times 0.99^2 \times 3^2 \times 10^{16}) = 1 \times 10^{16}$ Pa. You can see the big difference between these frames. The restricting pressure for the muon flying toward the surface is about $10^{11}$ higher $(10^{16}/10^5)$ than in the laboratory. The muon travels toward the surface at the speed of 0.99**c** in an environment that pushes against its tendency to emit neutrinos at a rate about 100,000 million times higher than in the laboratory.

If the force restricting decay has increased $10^{11}$ times, then the lifetime of the muon increases from $2.2 \times 10^{-6}$ sec to $2.2 \times 10^5$ seconds, or 2.5 days. This explains why we can easily register copious muons on the surface of the Earth; we don't have to change physics to account for it.

Let's try using a colliding force. The mass of the muon is $1.88 \times 10^{-28}$ kg, so the linear momentum of the muon is $5.6 \times 10^{-20}$ ($1.88 \times 10^{-28} \times 0.99c$). If this muon collides with the Earth, and this collision lasts for the lifetime of the muon, then this momentum is changed into impulse (force $\times$ colliding time). Then the colliding force is ($5.6 \times 10^{-20}/2.2 \times 10^{-6}$ sec) $= 2.5 \times 10^{-14}$ N. When the muon hits a barrier with this force, then the same force must act against the muon according to the Law of Action and Reaction.

The laboratory frame has an atmospheric pressure of $10^5$ Pa, and if the muon has the radius of 0.25 of the proton's radius, its radius is $0.22 \times 10^{-15}$ m. Thus, the surface area is $[4 \times 3.14 \times (0.22 \times 10^{-15})^2] = 0.6 \times 10^{-30}$ m$^2$. The atmospheric force on the surface is equal to pressure times area: ($10^5 \times 0.6 \times 10^{-30}$) $= 6 \times 10^{-26}$ N.

Thus, the impulse force or restricting force of the moving muon is $4.2 \times 10^{11}$ ($2.5 \times 10^{-14}/6 \times 10^{-26}$) higher than the force acting on the surface of the muon in the laboratory. This high colliding force means that muons can penetrate tens of meters into the Earth.

My assertion is that traditional physics already adequately explains the muon's unusual survival from the upper atmosphere to the Earth's surface, and therefore the longer decay time should not be used to prove strange theories, such as the idea that time itself changes according to different frames of reference. Physicists cannot ignore the existence of different reference frames, but in one frame of reference, the muon is at rest; and in the second, it's moving at almost the speed light through a medium that offers resistance. Therefore, only physicists who would take the laboratory frame and put it on a moving frame, like a rocket, can argue as to whether the muon's lifetime changes in relation to speed or not.

We need to stop misleading people—even school children—with the idea that time dilatation is proven by a muon's lifetime!

## Time and Mass

According to Einstein, time expands near a large mass and contracts away from it. Let's stay on the Earth. If, according to Einstein, the Earth bends space and therefore the space near the Earth is contracted, then space devours time. If time is missing near the Earth, then the Earth had to swallow that time; which suggests, based on Einstein's theory, that mass devours time. If the Earth liquidates time, then does time exist in the center of the Earth? Or do subatomic particles continue to spin even in the Earth's core? Do they spin differently in the core than they do on Mount Everest?

Einsteinian physicists measure time both near the Earth and far from the Earth by using clocks. It's true that a clock ticks more slowly near or at a massive object. Let's see why.

## Clock Deviation in Different Environments

Today, we use atomic clocks to more accurately measure time. Their "ticking" is based on the dynamics of the subatomic particles comprising an atom.

We have electrons orbiting a nucleus consisting of protons and neutrons. All particles are in motion, and therefore have their own momentum. Electrons have a momentum that arises from their mass and velocity. This momentum tends to cause electrons to move in straight-line paths. But the attractive between electrons and protons keeps them together, so the electron circles the nucleus. This state is quite stable, and can serve as very accurate "ticking" for the atomic clock. The spinning of charged particles creates magnetism, and therefore there exists an electronic oscillation to the frequency of an atomic transition. This ticking is very precise, since electrons have very brief orbital periods. For instance, in cesium, the surface electron orbits the atom 9,192,631,770 times per second.

In order for the cycle to remain constant, there must no change in the momenta of the subatomic particles. The cesium clock uses the vast separation, relatively speaking, between the nuclear spin and the electron spin with which it interacts. The nuclear radius is $6.1 \times 10^{-15}$ meter. The orbit of the outer electron is $3.34 \times 10^{-10}$ meters, almost 55,000 times larger.

The definition of the second is now tied to the frequency associated with the transition in cesium-133, and so the cesium beam atomic clock is widely used as a standard. Yet, we know that photons can interfere with the electrons on an atom's surface, which means they may interfere with the surface electrons of cesium. If such an interaction occurs, then changes in the surface electrons due to that interaction must also change the orbital frequency and, thus, the "ticking" of this cesium clock. In addition to photons interfering with the cesium electrons, there's also the interference of gravitons to take into account. Let's consider just the surface electrons, occupying the outer or 6s orbit, and the nucleus entering into interactions.

**Change #1: Intensity of Photons**

Photons have momenta, and they interact with electrons on the surfaces of atoms. Let's consider the interaction when the electron has the same direction of rotational motion as the photon. Before the collision, the electron rotates along a circular path with constant angular speed, and therefore has angular momentum (rotational momentum). The photon propagates along a straight line, and has linear momentum. The "sticky" collision creates an electron-photon couplet having the combined momenta of both particles. In theory, during that collision the photon adds its linear momentum to the electron and so pushes the electron away from the nucleus slightly. In special cases, the photon can push the electron so hard that it's pushed out of the atom and continues as a single particle: the photoelectric effect. This can result in an electrical current, as taken advantage of in solar power cells. However, photons usually don't have enough momentum to produce electricity, and therefore the collisions push the electrons only a slight distance. The resulting interacting couplet orbits at a higher orbit than before due to its increased energy. The new orbit has a greater circumference, and therefore its period of rotation should increase. The longer period of rotation lowers the orbital frequency, which throws off the "ticking" of an atomic clock.

This is why an atomic clock should measure a slightly different time due to a changing intensity of photons. More photons means the atomic clock is delayed. Since photons come mostly from the Sun, then the atomic clock should tick slower during the day than during the night. **An atomic clock slows down when the intensity of photons increases.**


**Change #2: Intensity of Gravitons**

Gravitons are strings moving at the speed of light, and have momenta like photons do. The difference is in the form by which they propagate. Photons propagate as transverse waves, gravitons as longitudinal waves. Gravitons interact with strings that vibrate in the same way as they do, which is how a spring does. These strings comprise all particles having mass, since mass results from the dynamics of these strings in restricted regions, such as in the volumes of the subatomic particles. When these strings travel through space, they're graviton strings; when they move in a restricted volume, they're mass's strings. When those strings meet, they interfere with each other due to their familiarity.

We also need to use the physics for elastic "sticky" collisions in these circumstances, to understand how gravitons affect protons, neutrons, and electrons. The collision effect is an impulse of a force lasting for a very short time. These impulses take place for a brief period when the string inside matter achieves speed light, interferes with the graviton and both accelerate farther toward the middle of a restricted route (an abscissa), ending when the interacting strings start to slow to the speed of light on the way from the middle to the other side. During this collision, the mass's string had an average speed higher than the speed of free gravitons, and **lower than it had without the interaction.** This is due to the application of the Law of Conservation of Momentum for this collision, and indicates that gravitons lower the speed of a mass's string. It means that this string needs more time to reach the other side. Therefore, if once they make a clock working on frequency of mass's strings, then stronger gravitational field will slow its ticking.

In an atom, the colliding particle is partly pulled in the direction from which the graviton arrived (per Newton's Law of Universal Gravitation). If the atomic clock's speed is based on the surface electron and on the nucleus (as in the cesium atomic clock), then the stronger gravity field pulls the nucleus more toward the source of gravitons than the 6s electron, and therefore the electron has a partly deformed orbit. That is, the electron's orbit is not a perfect circle, but rather an ellipse. The ellipse thus requires a longer orbit. This increases the electron's orbital period. If that period increases, then the frequency must drop a bit. That is, **the atomic clock slows down in stronger gravitational field**.

**A Moving Clock**

According to the special theory of relativity, a moving clock ticks slowly with respect to the observer's clock. That is, the clock, having linear momentum, ticks slowly in relation to the clock without linear momentum.

Let's assume the same intensity of photons and gravitons, and make a change in the motion of a clock in which subatomic particles are used to measure the "ticking." All collisions depend on the momenta of all the objects entering the collision impulse. When the object (the clock) moves, the protons, neutrons and electrons are all moving. The conservation of energy is an absolute law for collisions. If collisions do not produce heat energy, then the Law of Conservation of Momentum determines the resulting effect of collisions.

When the clock moves to meet the photons, the colliding effects depend on the momentum of the clock, and therefore the colliding effect of the moving clock must differ from the colliding effect of a stationary clock. The moving clock experiences photons striking the surface electrons of the atoms used to measure time. We saw before that photons push on the interfering electrons, jumping them to a higher orbit in a classic quantum leap. We also have to take into account the increased momenta of the atoms themselves as they collide with the photons.

This increased force pushes electrons farther from the nucleus. The higher orbit where they end up has a larger orbital radius, and therefore the electron's orbit is longer. If the additional force does not push the electron into a higher orbit, then it should at least deform the circular orbit into an ellipse. The new orbit, whether higher or elliptical, increases the period of rotation, and thus lowers the electron's orbital frequency. Hence, a clock traveling eastward toward the sun should slow down in comparison to a stationary clock.

(Notice that the density of photons also increases when moving in the direction of the sun, and therefore the clock moving toward the sun ticks slower due to the increased density of photons, *and* slower due to the clock itself moving clock toward the sun. The resulting decrease in speed is a sum of both effects.)

If a clock moves away from a source of photons, the photons provide less momentum to the interacting electrons than they do in the stationary clock, and therefore the resulting circular orbit of the interacting couplet is not as long as it is during the standard interaction. This causes a shortened orbital period, and so the orbital frequency increases. The overall result should be that a clock traveling away from the source of photons should tick faster than a stationary clock.

Researchers have tested clocks in motion. They flew cesium atomic clocks east and west around the Earth in commercial airliners, to compare the elapsed time against that of a clock that remained at the U.S. Naval Observatory. Hafele and Keating, in 1971, found that the flying clocks lost 59 nanoseconds during the eastward trip and gained 273 nanoseconds during the westward trip.

Secondly: If a clock moves to meet gravitons (i.e., drops toward the Earth) then the pull of the subatomic particles toward the earth is less, and thus gravitational deformation of the electron's orbit is less elliptical than that of a clock not moving in the gravitational field—so the electron's orbital rate decreases, while the frequency of oscillation increases. The atomic clock should tick faster as it moves in the direction of the center of the Earth. A clock dropping toward the earth also enters into a denser gravitational field; so the resulting effect is a sum of both effects. We add the effect due to the moving clock, and subtract the effect due to the higher density of gravitons.

When the clock moves away from the source of gravitons (say, in a moving rocket), then interacting gravitons deform the orbits of the electrons in this atomic clock more than those in the stationary clock. This increases the orbital period, decreasing the orbital frequency. Thus, a clock moving away from an object producing gravitons ticks slower than a stationary clock. Again, we need to take into consideration the sum of two effects, since by moving away from a source of gravitation we also decrease the density of gravitons.

In our society there exists a superficial and scientific approach to such results, almost a conspiratorial approach to them. You can see this in relation to how we consider clocks. Researchers have found that a clock ticks differently depending on whether it's moving or not. The superficial view here is that maybe time has changed in this situation, or that maybe (the prevailing approach) the clock ticks differently depending on its speed.

The scientific approach to these results is to study whether the physical parameters have impacted the clock processes. But there's also the conspiratorial approach, where some researchers use unexpected results in an attempt to uproot physics.

## SPACE Interacting With Mass

Einstein's General Theory of Relativity explains gravity using geometry. According to Einstein, space contracts near a mass and dilates away from it, while time dilates near a mass and contracts away from it. We've already found that time does not depend on space (gravitation). A clock just ticks differently in different gravitational fields. Now, let's look at space in their spacetime continuum.

It seems from Einstein's definition that space does not exist in the volume of a massive object, or that at least it's smaller, since mass displaces it. But objects cannot cause a distortion, since space exists everywhere—whether in the core of the Sun, inside the Earth for miners, or in the interior of atoms, where electrons (having a diameter around $10^{-11}$ m) orbit around a nucleus having a diameter about $10^{-15}$ m. Even individual subatomic strings have enough space to move inside the volumes of subatomic particles to give them mass.

Einsteinian physicists claim that the changing geometry of space causes nearby objects to be rolled (pushed down) toward the massive object. This should be because less-dense space is created by curvature. Thus, **objects "roll down" from a region of higher density to a region of lower density of space.**

But our space is three-dimensional. Einstein explains gravitation by using two-dimensional space, picturing it as two-dimensional net. The massive object deforms this net, and so this net becomes three-dimensional, creating a hole or "gravity well." However, following this logic, the net is located on all sides of a globe, since gravity is more or less equal on each spot on the Earth's surface. Therefore, if pushing creates a hole on one side of the planet, then pushing into this hole on the other side should annul the existence of any hole. There could be only a higher density of space *at* the massive object, at the Earth.

Also, their general statement that space contracts near objects of ordinary matter and dilates away from it means that space is denser **closer** to the object, and dilated and thinner farther away from the object. Since they build their theory on different densities of space based on its proximity to objects of varying mass, then a smaller object should move from denser to "thinner" space. And **thus, according to this theory, a massive object should push a nearby small object away from itself, due to the existing gradient in the density of space.** The tendency is the opposite.

Our Earth has mass of $6 \times 10^{24}$ kg. If the mass of the neutron is $1.67 \times 10^{-27}$ kg, then there are about $3.5 \times 10^{51}$ baryons in the Earth's volume. The volume of the neutron is $1.88 \times 10^{-44}$ m$^3$, so the volume of baryonic matter if Earth were a neutron star would be $6.58 \times 10^7$ m$^3$. Or, if the Earth's volume decreases by a coefficient of $10^{12}$ (there is such a volumetric rate in an atom), and since its volume is $1.083 \times 10^{21}$, then the earth could have a packed volume about $10^9$ m$^3$. Earth's baryonic volume is therefore a sphere with a diameter of less than 500 m. Decreasing this volume again by this factor to pack in the strings of mass results in the earth being packed into a volume of one liter (1 cubic decimeter).

Could such a small volume of mass distort space around the Earth enough that even the moon would feel it? Could such a small volume create such a great force that it attracts asteroids to the Earth?

Let's consider the Sun. They say that the Sun creates a deep hole in the fabric of spacetime; the planets feel this curvature of space, and literally roll toward the Sun. Yet gravitation is a force in physics, and now we have the geometry of space producing the force.

However, how could this geometric force be applied to planetary geometry, as it is when they are lined up in the direction of the Sun? The geometrical disturbance of space caused by the Sun must change at this time, due to interference of space waves. Are the planets shifting from orbital movement to linear movement during this configuration? Or, when we observe a solar eclipse, do we lost some of the gravitational force from the Sun? No, since explaining gravitation as being caused by bent space is physical foolishness. This foolishness would not have occurred if physicists had introduced physical analyses into this new theory. They used the one physical aspect of objects "rolling" down spacetime toward more massive objects, but this aspect lacks any physical analysis, and therefore cannot be considered anything but a foolish physical theory.

Consider, for instance, a cylinder having base **S** and height **h**, placed perpendicular to the surface of the Earth. Buoyancy at the surface depends on a gradient of densities of space. It is as if the cylinder were placed into two liquids, partly in the water and partly in the air. When we take into consideration that the density of space changes due to altitude, then forces or pressures acting on the sides of the cylinder are annulled, and therefore, the buoyancy depends on densities at the bottom surface (base) and at top surface of the cylinder. Yet how big a gradient of density could there be at the surface of the Earth, to give such a gravitational buoyancy (gravitational force) to objects above the surface that they accelerate downward by $10m/s^2$? And what about the gradient of the density of space on the surface of the Sun, when the Sun gives an acceleration of 274 m/s$^2$, or on a neutron star, where acceleration is more than $10^{12}$ m/s$^2$—more than a billion kilometers per second?

If we want to know this gradient of spatial density, we still need to know the height involved. For instance, for a buoyancy caused by the density of the air (and thus acting against gravitation), a gradient density of one meter of altitude is about 0.000075 of the density on the surface. Air density (and thus air buoyancy) works for a short distance where there is the air, but the gravitational force of the Earth acts much farther out—as I've already mentioned, it's felt by the Moon. Therefore, a gradient of density of space cannot be so large that it causes such a gravitational buoyancy.

So, in returning to our cylinder, it depends how high the cylinder is. When the height is large, the density gradient is higher, and gravitational buoyancy must be more significant than it would be at a lower height: for instance, one millimeter. *This is incompatible with observed gravitation.* Gravitation is very well studied and analyzed, having been codified into laws that state that gravitational force depends on the mass of an object. Neither the shape of that object (and thus its geometry) or movement or lack thereof affect its gravity.

You can see from this simple physical analysis that Einstein's gravitational concepts do not belong in physics. Yet here we are, not in the year 1915 but in the year 2015, and Einstein's ideas still rule. We live in the Atomic Age, and can measure even the tiniest distances. We have constructed huge laser interferometers and have not detected any disturbances in space. Now, consider the fact that if researchers had found bent space around the Earth on their interferometers, then it would be on the order of micrometers over large distances. Would such a gradient of curvature be enough to create the high gravitational effect the Earth has on us, to make us so heavy? No. Even 1% of this gradient is still not enough attract me to the Earth so strongly that I can't even jump a meter from the surface. If they haven't yet found gravity waves, then it's very expensive foolishness to keep looking for them.

Even Einstein's explanation of gravitation would have some portion of a physical possibility, due to registering disturbance of a medium by an object; still, it is only theory of local force. However, astronomers know that gravity is a far-reaching force issued from every object deep into the universe. For instance, the black hole in the center of our galaxy extends its gravitational reach to universal bodies 50,000 light years away from it. Therefore, the effect of disturbance through the medium of space must be excluded from possible theories of gravitation.

These scientists use even very small deviations from Newtonian physics to support their theory, and then say that it is very well proven. But their tendency to select crumbs over loafs convinces me that they're on the wrong track. Therefore I won't continue to look at all the crumbs that Newtonian physics has lost. Instead, I will look only at the deviation in the orbit of Mercury.

There's a 0.012 degree difference in the orbit of Mercury per century in relation to Newtonian physics; this is said to prove Einsteinian curvature of spacetime. Yet, now we have the new discovery that besides gravity, there exists also another attractive force in the universe, based on what theorists call "dark matter." Proponents of the existence of "dark matter" also point to this discrepancy in Newton's equations.

But Mercury is very close to the Sun, so there *should* be some variances in its orbit, since we're unable to see all the physical effects in play. The Sun's very strong gravity could disrupt Mercury's orbit. Mercury's surface experiences enormous temperature variations, ranging from -173° to 427 °C (-173 to 800° F) during the course of a Mercurial day. Since its day is longer than its year (indeed, an observer on Mercury would see one day every two Mercurial years), then the moment of its inertia may differ, because the density of matter in the planet changes constantly due to thermal expansion. Also Mercury probably doesn't have a homogenous density, because many chemical compounds come together to create a planet. There should be some discrepancy in density on the northern hemisphere of the planet, in relation to its southern hemisphere. Even its geometrical shape isn't ideal.

Different materials could also create different electronic loops through which magnetic particles like the neutrino can travel, and so a different pull may be applied on northern hemisphere than on the south. Then too, asteroid strikes may affect the axis of rotation as well. The point is, Mercury is not a good choice for proving the curvature of space, since there are some strange effects that make its day equal to 88 Earth days (so Mercury is a weak "flywheel"), and this makes its axis of rotation unstable.

## Conclusions

Humans are imperfect in relation to *light*. We perceive that everything runs away from us, and so we feel that we are detain by our material world. Our bodies are chained to the Earth, and our minds are delayed due to the need to process information, thus adding to our delays in perception. One way to overcome this is to live in faith that any faults we see are not within us. Some physicians use this human longing to prescribe placebos (medically ineffectual treatments for illnesses), and the yearning of the patient for the placebo to work actually helps some people.

Albert Einstein did something similar for physics. His theory describes a way to stay on track with light (at sixteen, he wondered what it would be like to travel along with a light ray). He advises that if humans board a rocket traveling at the speed of light, we will enter a time that is other than earthly. In this frame of reference, things seem normal; in an external frame of reference, relative time, measurements of length, and the like will appear different. Time itself slows down in the new frame of reference, as perceived from the outside. Time changed by high speed creates new activities in the new world/reference frame.

I have showed in this book that all these effects are illusions. The standard unit of time never changes, because it is rooted in the world of elementary strings. What changes is how long it takes us to process light. We have to pick up the input, deliver it into our minds, elaborate on it, manufacture it into output, and sometimes deliver it back to the outside world in order to take action. The cognitive conclusion is that when human mental processes perceive the information from a train that travels along with a light ray, then the period between acquiring input and delivering output is shortened. Therefore, human imperfection in perceiving nature seems taken away by the speed. Still, however, natural time is *not* affected by our limits in perception, so despite what we think we see, time dilation does not truly exist in nature.

Another of Einstein's prescriptions goes by the name of General Relativity. This drug attempts to cure human discomfort by explaining gravitation. He sees humans as caught in a net of curved spacetime. This net is active and wraps up those who enter it, jailing them. Since we long to be free, some of us have taken the first steps toward being free, which lies in understanding the fabric of this net. Thus they have accepted Einstein's placebo, and now walk in our world as higher beings because they have taken the first step to freedom, discovered the mystery of why the earth holds them back.

They omit the fact that Einstein's theory is not built on physical possibility, physical evidence and physical analysis. They believe that these crumbs of knowledge he has offered have cured them. Just as it's rare to find any healing substances in homeopathic drugs (since they are diluted by a factor $10^{6\text{-}12}$), it's is hard to find any serious physical evidence for the Einsteinian understanding of gravitation.

In reality, space is not curved and warped by celestial bodies. Since these theorists are unable to show any gravitational waves among objects in our world, or detect holes in real space, they look for gravitational waves in space left over by the Big Bang, hoping that will cure them.

Einstein's theories of relativity use the human cognitive process of *misunderstanding* human perception, and therefore there is no serious evidence for them. In this book, I believe that I have uprooted all the main evidence for Einsteinian theories built on dilation of time and space. If not, let me know, and I will look into that evidence as well—and together, we can cleanse the human "odor" from our science.

## The End

# Appendix

| The Human Odor – Quantum Physics | The Reality - String Clasical Physics |
|---|---|
| Special theory of relativity (SR) is a physical theory, since speed changes time and length | It is a philosophical theory, since shows how we see world due our delayed perception of light |
| The speed of alien photons is the same as ours, – confirms SR | Photon speed is based on the local light field, which forces all photons to travel at the same speed. Ours is c |
| Lifetime of moving muons confirms time dilation in SR | Air drag does not allow muons to decay |
| The speed of light in a moving transparent medium is the same as in a stationary transparent medium – SR | Absorption and re-emission light in a medium depends on the density of the medium, not its speed |
| A clock ticks differently moving eastward and westward in relation to a stationary clock   SR | The law of momentum conservation for subatomic particles of a clock and photons is at work here |
| The General Theory of Relativity (GR) is physical theory for gravitational force (the Sun causes a distortion in the fabric of space) | It does not, because we do not have physical analyses on whether wave and viscous effects caused by a disturbance in a medium can be stretched 11 billion miles out |
| A clock ticks slower in a stronger gravitational field – GR | Interacting gravitons slow the motion of subatomic particles |
| Space contracts near a mass and dilates away from it – GR | Observed gravitation in the far universe does not allow for this |
| Lensing of light is due to gravitation – GR | Refraction due to different transmission media |
| Quarks are fundamental constituents of baryons – SM | Strings are the fundamental particles and come from matter |
| Three quarks make up a baryon – SM (Standard Model) | Quarks make up less than two percents of a baryon's mass |
| Forces of quark bonds provide over 98% mass to baryons - SM | Force is the interacting effect of objects, force is not mass |
| Quarks have electric potentials of "up" +2/3; "down"-1/3 – SM | Diferent electrical potentials are short-circuited |
| Quarks cannot be squeezed, since they are fundamental constituents of matter; thus baryons cannot be squeezed – SM | They're squeezed in a black hole, and protons are squeezed into the extreme density in LHC during a collision |
| Mass is the interacting effect of subatomic particles with alien particles (Higgs bosons) – SM | Dynamics of strings in subatomic particles create mass; they move there, since there is plenty of free space for movement |
| Gravitation is an effect of the curvature of space - GR versus ST | Gravitons interfere with mass's strings in subatomic particles |
| Elementary negative electric potential is a scalar one-dimensional physical quantity | A physical spinning string with one end creating a vortex, a three dimensional dynamic object, creates this potential |
| A magnetic monopole is a isolated magnet | It is a single looped string |
| A neutrino is electrically neutral and therefore is not affected by the electromagnetic force | A neutrino is magnetic, and therefore electric particles (electron and proton) deviate it from themselves |
| Strong nuclear force is interacting gluons between quarks | Contacting strings holds others inside their vortices |
| Elementary entropy is a degree of disorder or randomness in the system | It is the energy that binds strings in subatomic particles, and vanishes during spontaneous emission of strings |
| Photons create lines for static magnetic and electric field | They are created by free strings without linear momentum |
| Dark matter is responsible for additional attraction | Neutrinos interacting with electronic fields cause the attraction |
| Dark energy is responsible for pushing objects in the universe | It is caused by the linear momentum of photons |
| Abstract quantum mechanics and chromodynamics | Abstracts cannot be use for concrete objects in physics |
| Our world is 26-dimensional, or if supersymmetry exists, then it is 11-dimensional | Any signals of motion and life from extra dimensions do not enter into our three-dimensional world |
| Mathematical theories rule physics, by a model built from abstractions (quantum field theory), probabilities, symmetries, geometry and other mathematical elements | The role of physics is to cleanse the theory of nature of abstracts, probabilities, and myths; concrete physical objects are the real carriers of physical forces |

www.ingramcontent.com/pod-product-compliance
Lightning Source LLC
Chambersburg PA
CBHW050734180526
45159CB00003B/1224